# Thermoelectric Materials Research and Device Development for Power Conversion and Refrigeration

T0351279

MATERIALS RESEARCH SOCIETY
SYMPOSIUM PROCEEDINGS VOLUME 1490

# Thermoelectric Materials Research and Device Development for Power Conversion and Refrigeration

Symposium held November 25–30, 2012, Boston, Massachusetts, U.S.A.

**EDITORS**

## George S. Nolas
*University of South Florida*
Tampa, Florida, U.S.A

## Yuri Grin
*Max-Planck Institute for
Chemical Physics of Solids*
Dresden, Germany

## David Johnson
*University of Oregon*
Eugene, Oregon, U.S.A.

## Alan Thompson
*Marlow Industries, Inc.*
Dallas, Texas, U.S.A.

Materials Research Society
Warrendale, Pennsylvania

CAMBRIDGE
UNIVERSITY PRESS

# CAMBRIDGE
## UNIVERSITY PRESS

Shaftesbury Road, Cambridge CB2 8EA, United Kingdom

One Liberty Plaza, 20th Floor, New York, NY 10006, USA

477 Williamstown Road, Port Melbourne, VIC 3207, Australia

314–321, 3rd Floor, Plot 3, Splendor Forum, Jasola District Centre, New Delhi – 110025, India

103 Penang Road, #05–06/07, Visioncrest Commercial, Singapore 238467

Cambridge University Press is part of Cambridge University Press & Assessment, a department of the University of Cambridge.

We share the University's mission to contribute to society through the pursuit of education, learning and research at the highest international levels of excellence.

www.cambridge.org
Information on this title: www.cambridge.org/9781605114675

Materials Research Society
506 Keystone Drive, Warrendale, PA 15086
http://www.mrs.org

First published 2013

CODEN: MRSPDH

*A catalogue record for this publication is available from the British Library*

ISBN    978-1-605-11467-5    Hardback

# CONTENTS

*Invited Paper

v

NANOCOMPOSITES AND NANOSTRUCTURED MATERIALS

*Invited Paper

## THERMOELECTRIC PROPERTIES AND APPLICATIONS

# THIN-FILMS

*Invited Paper

ix

# PREFACE

Symposium B, "Thermoelectric Materials Research and Device Development for Power Conversion and Refrigeration," held from November 26 – 30 at the 2012 MRS Fall Meeting in Boston, Massachusetts, was the tenth in a series of symposia on state-of-the-art materials and technologies for direct thermal-to-electric energy conversion that produced proceedings with primary focus on material and technological advances of thermoelectrics and thermionics [see MRS Proceedings volumes 234, 478, 545, 626, 691, 793, 886, 1044 and 1166]. In this symposium there were 263 contributed presentations, the largest by far at the MRS, including 10 invited talks and 168 poster presentations. These presentations were given from researchers from academia, national laboratories, and industry in the United States, Asia and Europe. The symposia covered a broad range of topics in the areas of materials, measurement techniques, and device development.

The papers were divided into different sections and included Bulk Materials, such as Skutterudites, Clathrates, different chalcogenide compositions and oxides, Thin Films, Thermionics, Nanocomposites and Nanostructured Materials and Device Development. This volume provides an overview of the exciting recent developments in the field. The continuing interest in the area of thermoelectric materials research, as well as device-related research, was evidenced by the excellent level of attendance throughout the symposium.

As with previous symposia in this series, there were a large number of graduate student presentations. Such student participation continues to be a focus of our symposium, emphasizing the mentoring of our future scientists in this field of materials research. With the generous support from our sponsors, the symposium organizers were able to give eighteen student presentation awards.

Oral Presentations

Samantha Clarke (University California, Los Angeles), "Optimizing Thermoelectric Efficiency of $La_{3-x}Te_4$ with Alkaline Earth Metal Substitution"

Jeffrey Doak (Northwestern University), "Dopants, Solubility, and Vibrations in PbS-PbTe Alloys"

Rachel Korkosz (Northwestern University), "High Performance Na-doped PbTe-PbSe-PbS Thermoelectric Materials"

Jaeho Lee (Stanford University), "Phase Purity and the Thermoelectric Properties of $Ge_2Sb_2Te_5$ Films for Phase Change Memory"

Gloria Lehr (Michigan State University), "A Study of the Relationship between Intermediate Valence and Seebeck Coefficient in Yb-Al Compounds"

Pooja Puneet (Clemson University), "Decoupling Thermoelectric Properties of Polycrystalline Bi via Surface Modification"

Poster Presentations

Peng Gao (Michigan State University), "Optimizing the Thermoelectric Properties of $Mg_2Si_{0.4}Sn_{0.6}$ by Mg Compensation and Sb Doping"

Shintarou Mikami (Muroran Institute of Technology, Japan), "Sintering of $Cu_2ZnSnS_4$ Synthesized by Solid-phase Reaction of Sulfides and its Thermoelectric Properties"

Maribel Maldonado (University of Texas at Dallas), "Deposition and Doping of Thin-film $Mg_2Si$"

Hua He (University of Delaware), "Synthesis, Crystal Structure, Electronic Structure, and Thermoelectric Properties of $Rb_2Cd_5As_4$ and the Solid Solution $Rb_2Cd_5(As,Sb)_4$"

Mohsin Saleemi (KTH Royal Institute of Technology, Sweden), "Fabrication of Nanostructured Bulk Cobalt Antimonide ($CoSb_3$) Based Skutterudites via Bottom-up Synthesis"

Bolin Liao (Massachusetts Institute of Technology), "Cloaking Core-shell Nanoparticles from Conducting Electrons in Solids"

Laetitia Boulat (University of Montpellier, France), "Diffusion Barriers for $CeFe_4Sb_{12}/Cu$ Thermoelectric Devices"

Michael Gaultois (University of California, Santa Barbara), "Exploiting Electronic Instability to Optimize Seebeck and Electrical Conductivity"

The organizers are most grateful for the support of the Aldrich Materials Science, FCT Systeme GmbH, Fuji Electronic Industrial Co, Ltd., GE Global Research, General Motors Corp., Marlow Industries, Inc., a subsidiary of II-VI Inc., M. Braun, Inc., and Thermal Technologies, LLC. This support enabled the eighteen student awards.

G.S. Nolas
Y. Grin
D. Johnson
A. Thompson

# MATERIALS RESEARCH SOCIETY SYMPOSIUM PROCEEDINGS

# MATERIALS RESEARCH SOCIETY SYMPOSIUM PROCEEDINGS

# MATERIALS RESEARCH SOCIETY SYMPOSIUM PROCEEDINGS

Volume 1534E — Low-Dimensional Semiconductor Structures, 2012, T. Torchyn, Y. Vorobie, Z. Horvath, ISBN 978-1-60511-511-5

Prior Materials Research Society Symposium Proceedings available by contacting Materials Research Society

**Bulk Thermoelectrics**

Mater. Res. Soc. Symp. Proc. Vol. 1490 © 2013 Materials Research Society
DOI: 10.1557/opl.2013.85

# High-temperature thermoelectric properties of W-substituted CaMnO$_3$

Dimas S. Alfaruq [a], James Eilertsen [a], Philipp Thiel [a], Myriam H Aguirre [a], Eugenio Otal [a], Sascha Populoh [a], Songhak Yoon [a], Anke Weidenkaff [a,*]

[a] Empa, Solid State Chemistry and Catalysis, Ueberlandstrasse. 129, CH-8600 Duebendorf, Switzerland

*Corresponding Author

**Abstract**

The thermoelectric properties of W-substituted CaMn$_{1-x}$W$_x$O$_{3-\delta}$ (x = 0.01, 0.03; 0.05) samples, prepared by soft chemistry, were investigated from 300 K to 1000 K and compared to Nb-substituted CaMn$_{0.98}$Nb$_{0.02}$O$_{3-\delta}$. All compositions exhibit both an increase in absolute Seebeck coefficient and electrical resistivity with temperature. Moreover, compared to the Nb-substituted sample, the thermal conductivity of the W-substituted samples was strongly reduced. This reduction is attributed to the nearly two times greater mass of tungsten. Consequently, a ZT of 0.19 was found in CaMn$_{0.97}$W$_{0.03}$O$_{3-\delta}$ at 1000 K, which was larger than ZT exhibited by the 2% Nb-doped sample.

## 1. Introduction

Thermoelectric materials convert heat into electricity directly. Their thermoelectric efficiency is nominally quantified by a dimensionless figure of merit (ZT). The ZT is a function of the interdependent electrical and thermal transport properties of the material; specifically, the Seebeck coefficient (S), electrical resistivity ($\rho$), and total thermal conductivity ($\kappa$) – determined at a particular temperature (T) in Kelvin: $ZT = S^2 T/\rho\kappa$, where $\kappa = \kappa_l + \kappa_e$.

Consequently, in order to enhance the ZT of a thermoelectric material, a high Seebeck coefficient must be maintained while the electrical resistivity and the total thermal conductivity of the material are reduced[1]. This can be achieved, in part, by introducing additional charge carriers up to a suitable concentration. As a result, the electrical resistivity can be reduced without reducing the Seebeck coefficient appreciably. Aliovalent transition-metal cation substitution, in particular with comparatively heavy transition metals, can achieve both increased charge carrier concentration and decreased lattice thermal conductivity due to mass-difference phonon scattering [1-3].

The calcium manganate, CaMnO$_3$, is a promising thermoelectric material as it exhibits high Seebeck coefficients, strong resistance to chemical degradation at elevated temperatures, and a perovskite-type structure susceptible to A- and B-site substitution [4]. Therefore, this paper reports an investigation into the effect of aliovalent B-site substitution on the structure and thermoelectric properties of CaMnO$_3$. W-substituted CaMn$_{1-x}$W$_x$O$_{3\pm\delta}$ (x = 0.01; 0.03; 0.05) samples were prepared by soft chemistry, sintered, and characterized by x-ray diffraction and electron microscopy. Thermogravimetric analysis experiments were performed to determine the oxygen stoichiometry of each sample. The thermoelectric properties were measured from 300 K to 1000 K and compared to Nb-substituted CaMn$_{0.98}$Nb$_{0.02}$O$_{3-\delta}$[4] in order to evaluate the effectiveness of W-substitution.

## 2. Experimental

Polycrystalline perovskite-type CaMn$_{1-x}$W$_x$O$_{3\pm\delta}$ (x = 0.01; 0.03; 0.05) samples were synthesized via a procedure developed by Bocher et al.[4]. The samples were prepared by dissolving stoichiometric amounts of Ca(NO$_3$)$_2$·4H$_2$O (Fluka, ≥99.0%), Mn(NO$_3$)$_2$·4H$_2$O (Merck, 98%), and citric acid (Alfa Aesar, +98%) in distilled water. The WCl$_6$ (Aldrich, 99.0%) was dissolved in isopropanol and added to the citric acid solution. The citric acid to total cation (Ca, Mn, W) ratio was 2:1. The solution was polymerized at 353 - 363 K for 4 hours under continuous stirring. The polymerized product was heated at 353 K for 16 hours to obtain the xerogel precursor. The xerogel precursor was heated to 573 K at a rate of 20 K/min to elimi- nate organic material; then crushed, ground in a mortar and pestle, heated at a rate of 5 K/min to 1373 K, and calcined for 8 hours. The Nb-substituted $CaMn_{0.98}Nb_{0.02}O_{3\pm\delta}$ sample was syn- thesized as above. The samples were pressed uniaxially using a hydrostatic press, heated at a rate of 5 K/min to 1473 K, sintered for 5 h, and cooled at a rate of 5 K/min.

Powder X-ray diffraction (PXRD) patterns were obtained using a PANalytical X'Pert PRO $\theta$-$2\theta$ scan system equipped with Johansson monochromator and the ultra-fast X'Celera- tor linear detector. The incident X-rays had a wavelength of 0.154056 nm (Cu-K$\alpha_1$). The dif- fraction patterns were scanned from 20° to 100° ($2\theta$) with an angular step interval of 0.0167°. PXRD data was analyzed with the Le Bail technique by using the Fullprof Suite software package [5,6], and Thompson-Cox-Hastings pseudo-Voigt profile fitting functions. High- resolution transmission electron microscopy (HRTEM) and electron diffraction (ED) studies were carried out using a Jeol JEM FS2200 electron microscope.

The oxygen stoichiometry was determined by reducing the sintered samples and measuring their mass loss using a Netzsch STA 409 CD thermogravimetric analysis (TGA). The samples were heated at a rate of 20 K/min to 1373 K under a reducing atmosphere (20% vol $H_2$/He). The atmosphere reduced the $CaMn_{1-x}W_xO_{3\pm\delta}$ phase and generated an evolution of $H_2O$ vapor and the formation of MnO and CaO:

$$CaMn_{1-x}W_xO_3(s) + H_2 \xrightarrow{1373K} CaO(s) + (1-x)MnO(s) + (x)W(s) + H_2O \ (g) \quad \text{(eq.2)}$$

Thermoelectric data were collected from 300 K to 1000 K in synthetic air on samples sintered to approximately 85% of their theoretical density. Seebeck coefficient (S) and elec- trical resistivity ($\rho$) data were collected on rectangular-shaped samples by using an RZ2001i unit from Ozawa Science (Japan). Thermal diffusivities were measured using a Netzsch LFA 457 laser flash apparatus. A Netzsch DSC 404 C Pegasus with sapphire standard was used to measure the heat capacity ($C_p$) of the compounds by the ratio method. Thermal conductivity ($\kappa$) was determined from thermal diffusivity ($\alpha$) and heat capacity ($C_p$) data, and the sample density ($d$) according to the relationship: $\kappa = \alpha^* d^* C_p$.

## 3. Results and Discussions

PXRD data and the Le Bail profile fit of the $CaMn_{0.99}W_{0.01}O_{3-\delta}$ sample are shown (Fig. 1a). All samples crystallized into the perovskite-type orthorhombic crystal structure with *Pnma* space group. A systematic increase in the unit cell volume is observed with tungsten substitution (Fig.1b). The unit cell volume data[7] for $CaMnO_3$ is included in the figure. The tungsten-substituted samples do not deviate from linearity appreciably. Note the perovskite- type crystal structure depicted in the inset.

Twin domains were observed by using HRTEM (Fig. 2). The Fast Fourier Trans- formed (simulated by Digital Micrograph 3.8.2 Gatan software) taken from different regions of the HRTEM image indicates the superposition of the two zone-axis ([10-1] and [101]), the- se are identical in the orthorhombic structure and form twin domains. These domains are typi- cally between 30 and 50 nm. The twin domains likely occur because the $CaMnO_3$ compound undergoes a structural transition from a higher symmetry, cubic structure (the high- temperature structure), to a lower symmetry, orthorhombic structure (the low-temperature structure). Since the deviations from the cubic symmetry are small and the orthorhombic structure of $CaMnO_3$ can be described with $\sqrt{2}$a x 2a x $\sqrt{2}$a unit cell, domains with different unit-cell orientations can grow coherently, thus yielding the observed twinning phenomenon[8,9].

Oxygen deficiencies of $\delta$=0.039, 0.112, and 0.179 for the respective $x$ = 0.01, 0.03, and 0.05 $CaMn_{1-x}W_xO_{3-\delta}$ samples were determined by TGA. A systematic increase in oxygen deficiency occurs with increasing tungsten concentration.

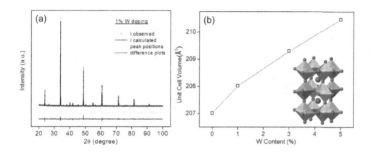

Figure 1: PXRD data and Le Bail profile fit of $CaMn_{0.99}W_{0.01}O_{3\pm\delta}$ (a). The unit cell volume expansion with increasing tungsten concentration (b). The crystal structure of the calcium manganate perovskite is shown (inset): the red spheres represent calcium, the blue represent oxygen, and the manganese and tungsten are located at the center of the octahedra.

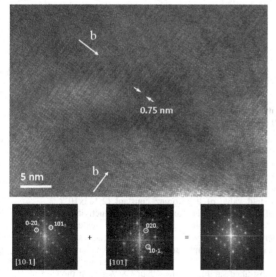

Figure 2: HRTEM of $CaMn_{1-x}W_xO_3$ showing a twin domain structure (the [10-1] and [101] zone axis).

5

The electrical resistivity data is shown (Fig. 3a). The electrical resistivity of all samples increases with increasing temperature, a trend consistent with heavily doped semiconductors. Moreover, the electrical resistivity drops with increasing tungsten concentration. The resistivity of the 2% Nb-substituted sample is higher than all of the W-substituted samples over the majority of the temperature range measured. Aliovalent substitutions of both $W^{6+}$ and $Nb^{5+}$ for $Mn^{4+}$ likely induce mixed valence cations; namely, $Mn^{3+}$ and $Mn^{4+}$. Electrons can hop from the $Mn^{3+}$ to $Mn^{4+}$, mediated by the 2p electrons in oxygen, thus decreasing the electrical resistivity[10]. Therefore, tungsten will supply more charge carriers than an equivalent amount of niobium, providing a possible explanation for the comparatively lower electrical resistivity in the W-substituted samples [11]. The electrical resistivity of the 1% W- and 2% Nb-substituted samples, however, are quite similar. Since tungsten and niobium likely possess hexavalent and pentavalent oxidation states, respectively, the concentration of $Mn^{3+}$ in both compositions is likely to be similar – leading to similar electrical resistivity in both.

The Seebeck (S) coefficient data is shown (Fig. 3b). Sizable negative Seebeck coefficients are observed in all samples. The magnitude of the Seebeck coefficients increased with increasing temperature consistent with the concomitant rise in electrical resistivity. Moreover, the magnitude of the Seebeck coeffcients tended to drop with increasing tungsten substitution as the charge carrier concentration increases (and the electrical resistivity decreases) with substitution [1,12,13].

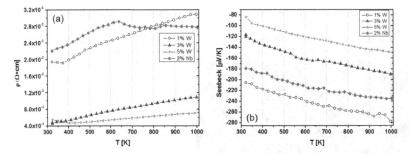

Figure 3: Temperature dependent resistivity (a) and Seebeck coefficient (b) data of the W-substituted $CaMn_{1-x}W_xO_3$ (x = 0.01; 0.03; 0.05) and 2% Nb-substituted samples.

The total thermal conductivity and lattice thermal conductivity data of the W-substituted $CaMn_{1-x}W_xO_{3\pm\delta}$ (x = 0.01; 0.03; 0.05) and 2% Nb-substituted $CaMn_{0.98}Nb_{0.02}O_{3\pm\delta}$ samples are shown (Fig. 4a, b). The lattice thermal conductivity is calculated by subtracting the electronic component from the total thermal conductivity. The electronic component was calculated from the electrical resistivity data using the Wiedemann-Franz law: $\kappa_e = L_oT/\rho$, where $L_o$ was taken as the Sommerfeld value of the Lorenz constant ($2.443 \times 10^{-8}$ $W\Omega/K$), and $\rho$ is the electrical resistivity of the material. The total thermal conductivity is dominated by the lattice component as observed in other substituted calcium manganates[3,4,13,14]. The total and lattice thermal conductivity of the 2% Nb-substituted sample is higher than all of the W-substituted samples. For example, at 1000 K the lattice thermal conductivity of the 2% Nb-substituted sample reached 2.16 W/mK while all of the W-substituted samples, including the 1% W-substituted sample, are less than 1.8 W/mK at the same temperature. Since the atomic mass of tungsten is nearly two times that of niobium, and more than three times the mass of manganese, tungsten substitution enhances the mass-difference phonon scattering in the crystal structure significantly. Moreover, at lower temperatures, the lattice thermal conductivity is reduced more significantly as mass-difference phonon scattering is more effective in this temperature range.

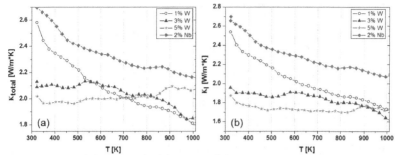

Figure 4: (a) Total thermal conductivity ($\kappa_{total}$) of $CaMn_{1-x}W_xO_3$ (x= 0.01; 0.03; 0.05) and $CaMn_{0.98}Nb_{0.02}O_{3\pm\delta}$; (b) Electronic thermal conductivity ($\kappa_{el}$) of $CaMn_{1-x}W_xO_3$ (x= 0.01; 0.03; 0.05); (c) Phonon thermal conductivity ($\kappa_{ph}$) of $CaMn_{1-x}W_xO_3$ (x= 0.01; 0.03; 0.05)

The ZT of the $CaMn_{1-x}W_xO_{3\pm\delta}$ (x = 0.01; 0.03; 0.05) and 2% Nb-substituted $CaMn_{0.98}Nb_{0.02}O_{3\pm\delta}$ sample was calculated for the 300 K to 1000 K temperature range. The 3% W-substituted $CaMn_{0.97}W_{0.03}O_{3\pm\delta}$ sample exhibits the highest ZT of the W-substituted samples, nearly reaching 0.2 at 1000 K. The low electrical resistivity and thermal conductivity achieved by tungsten substitution enhances the ZT significantly. However, the ZT of 2% Nb-substituted nanostructured samples reached 0.32 at 1060 K[4].

## 4. Conclusion

Polycrystalline W-substituted $CaMn_{1-x}W_xO_{3\pm\delta}$ (x = 0.01; 0.03; 0.05) samples were synthesized by a soft chemistry method. All samples crystallized into an orthorhombic crystal structure with *Pnma* space group. It was found that there was a systematic decrease in oxygen content with increasing tungsten concentration. The Seebeck coefficient, electrical resistivity, and thermal conductivity data were measured at elevated temperatures. The Seebeck coefficient and electrical resistivity is reduced with increasing tungsten substitution. The lowest electrical resistivity was achieved by the $CaMn_{0.95}W_{0.05}O_{3\pm\delta}$ sample while the $CaMn_{0.99}W_{0.01}O_{3\pm\delta}$ sample exhibits the highest Seebeck coefficient value at 1000 K (-270 $\mu$V/K). The thermal conductivity of all of the W-substituted samples is reduced in comparison to the 2% Nb-substituted sample due to enhanced mass-difference phonon scattering. Due to the interdependence of the Seebeck coefficient and the electrical resistivity, the best thermoelectric performance of the tungsten-substituted samples was obtained in the 3% W-substituted $CaMn_{0.97}W_{0.03}O_{3\pm\delta}$ sample with a ZT of 0.19 at 1000 K. The ZT was found to be lower than the 2% Nb-substituted sample, however, which reached a ZT of 0.32 at 1060 K[4]. Nevertheless, tungsten substitution is more effective than niobium substitution at reducing the thermal conductivity of the calcium manganates.

## Acknowledgements

The authors would like to thank the Swiss Federal Office of Energy (BfE), Sinergia TEO (SNF) and Empa for the financial support.

# References

(1) Rowe, D. M. *Thermoelectrics Handbook - Macro to Nano*; CRC Press/Taylor & Francis Group: Boca Raton, 2006.

(2) Funahashi, R.; Kosuga, A.; Miyasou, N.; Takeuchi, E.; Urata, S.; Lee, K.; Ohta, H.; Koumoto, K. *Appl. Phys. Lett* **2007**, p 124.

(3) Wang, Y.; Sui, Y.; Wang, X.; Su, W.; Liu, X.; Fan, H. J. *Acta Mater.* **2010**, *58*, 6306.

(4) Bocher, L.; Aguirre, M. H.; Logvinovich, D.; Shkabko, A.; Robert, R.; Trottmann, M.; Weidenkaff, A. *Inorg. Chem.* **2008**, *47*, 8077.

(5) Lebail, A.; Duroy, H.; Fourquet, J. L. *Mater. Res. Bull.* **1988**, *23*.

(6) Rodriguez-Carvajal, J. *Physica B* **1993**, *192*, 55.

(7) Poeppelmeier, K. R.; Leonowicz, M. E.; Scanlon, J. C.; Longo, J. M.; Yelon, W. B. *Journal of Solid State Chem.* **1982**, *45*.

(8) Aguirre, M. H.; Canulescu, S.; Robert, R.; Homazava, N.; Logvinovich, D.; Bocher, L.; Lippert, T.; Dobeli, M.; Weidenkaff, A. *J. Appl. Phys.* **2008**, *103*, 013703.

(9) M.H. Aguirre, D. L., L. Bocher, R. Robert, S.G. Ebbinghaus and A. Weidenkaff *Acta Mater.* **2008** *57*, 108.

(10) Zener, C. *Phys. Rev.* **1951**, *82*, 403.

(11) Raveau, B.; Zhao, Y. M.; Martin, C.; Hervieu, M.; Maignan, A. *J. Solid State Chem.* **2000**, *149*, 203.

(12) Horiguchi, K. I.; Teduka, Y.; Sugihara, S. *Funtai Oyobi Fummatsu Yakin/ J. Japan Soc. of Powder and Powder Metall.* **2007**, *54*, 351.

(13) Wang, Y.; Sui, Y.; Fan, H.; Wang, X.; Su, Y.; Su, W.; Liu, X. *Chem.Mater.* **2009**, *21*, 4653.

(14) Maignan, A.; Martin, C.; Autret, C.; Hervieu, M.; Raveau, B.; Hejtmanek, J. *J. Mater. Chem.* **2002**, *12*, 1806.

Mater. Res. Soc. Symp. Proc. Vol. 1490 © 2013 Materials Research Society
DOI: 10.1557/opl.2013.216

# Electron and Phonon Transport in n- and p-type Skutterudites

Jiong Yang[1], S. Wang[1], Jihui Yang[1, a)], W. Zhang[2], and L. Chen[2]

[1]Materials Science and Engineering Department, University of Washington, Seattle, WA
98195-2120, USA
[2]State Key Laboratory of High Performance Ceramics and Superfine Microstructure,
Shanghai Institute of Ceramics, Chinese Academy of Sciences, Shanghai 200050, China

ABSTRACT

Filled skutterudites are one of the most promising materials for thermoelectric (TE) power generation applications at intermediate temperatures due to their superior TE and thermomechanical performance as compared to other materials. In the past, we have demonstrated that n-type skutterudites can be optimized so that their maximum TE figure of merit reaches 1.7 at 850 K. TE performance of the p-type, however, is lagging behind, which hinders the optimization of skutterudites-based TE module development. In this paper we reveal that the underlying reasons for inferior TE properties of the p-type root in their electronic band structures, which result in higher thermal conductivity at elevated temperatures due to bipolar lattice thermal conduction and lower power factor because of heavy valance bands induced strong electron-phonon interactions. We also identify means of improving the power factor and reducing bipolar effect.

a) Author to whom correspondence should be addressed, Electronic mail: jihuiy@uw.edu

## INTRODUCTION

One of the key challenges on thermoelectric (TE) research is to search for high efficiency TE materials, which should possess high thermopower ($\alpha$), high electrical conductivity ($\sigma$), low thermal conductivity ($\kappa$ - summation of the electronic $\kappa_e$, lattice $\kappa_L$, and bipolar $\kappa_b$ components), and therefore high dimensionless TE figure of merit $ZT$ ($=\alpha^2\sigma T/\kappa$). Identifying materials with high $ZT$ values has proven to be extremely difficult since these transport properties are correlated with each other: increasing the thermopower usually means lowering the electrical conductivity, and vice versa; the electronic thermal conductivity also relates to the electrical conductivity via the Wiedemann-Franz law. Usually power factor (PF), $\alpha^2\sigma$, is determined by the electronic band structures (valence band (VB) or conduction band (CB)), and can be optimized as a function of carrier concentration and scattering mechanisms. On the other hand, the lattice thermal conductivity $\kappa_L$ is mainly determined by lattice dynamics and phonon scatterings. The $ZT$s in bulk materials have been improved significantly since 1990s with the majority of maximum values lower than 2.0, still not high enough to meet the need of large-scale industrial applications.[1]

Skutterudite compounds have been considered as promising candidates for advanced TE applications. Binary skutterudites, in the general formula of $MX_3$ (where M is Co, Rh, or Ir, and X is a pnicogen atom) crystallize in a body-centered-cubic structure with the space group $Im3$. The structure of $MX_3$ is composed of corner-sharing $MX_6$ octahedra.[2]    $CoSb_3$ is the most widely studied binary skutterudite in TE field. It exhibits high carrier mobility with relatively large effective mass, leading to reasonably high PFs. The structure of $CoSb_3$, as well as other binary skutterudites, contains large lattice voids which can be filled by guest atoms. Our earlier work on the maximum filling fraction, $i.e.$, the filling fraction limit for filler atoms in $CoSb_3$ led to the establishment of the so-called electronegativity rule for identifying fillers that could form stable filled $CoSb_3$.[3]    The results showed that only certain fillers could achieve stable partial filling, including rare earth (RE), alkali earth (AE), and alkali metal (AM) atoms.[4-10] It was also clarified that the interactions between the fillers and the neighboring Sb atoms are mostly Coulumbic and depend critically on the effective valence charge states of the fillers. RE, AE, and AM fillers show different valence charge states when filling into the lattice voids, but they all donate their outmost valence electrons to the host and contribute to the increase of carrier concentration.[11-13]

Thermoelectric properties of n-type filled $CoSb_3$ skutterudites are to a large extent related to their unique features of filling. $Ab$ $initio$ calculations have shown that the electronic states of fillers are at high energy region; the conduction band edge of filled $CoSb_3$, critical to electrical transport, remains unchanged by void filling.[11,13] It has been proven by many groups that the transport parameters of filled $CoSb_3$ skutterudites, such as thermopower and electrical conductivity, follow a similar trend versus the carrier concentration, determined by the Co-Sb framework. The optimal carrier concentrations are around 0.4~0.6 electrons per unit cell for maximizing PFs, regardless of the type or the number of filler species; they are slightly lower for optimizing ZTs. The maximum PFs with the optimal carrier concentrations in n-type $CoSb_3$ skutterudites can be over 50 $\mu$W/cm-$K^2$.[14-17]

Since filler atoms are loosely bound to the host, the atomic displacement parameters[18] of fillers are relatively large in comparison with those of Co or Sb. This feature leads to localized Einstein-like vibrational modes[19, 20] that can strongly scatter phonons and reduce the lattice thermal conductivity. Under a harmonic approximation, the resonant frequencies of fillers can be obtained by calculating the total energy as a function of the atomic displacement from their equilibrium positions.[21] Based on these results, filler atoms are categorized into three groups - AM, AE, and RE. The RE fillers have the lowest resonant frequencies as compared with the other groups, while Yb possesses the lowest amongst the REs.[21] Since $\kappa_L$ is largely contributed by the low frequency phonons, fillers with lower resonant frequencies would reduce $\kappa_L$ more effectively. Mixing fillers from different groups leads to a boarder spectrum of phonon scatterings and causes $\kappa_L$ approaching the minimal thermal conductivity at elevated temperatures.[22] Based on the aforementioned strategies, the optimized multiple-filled $CoSb_3$ skutterudites show a unique possibility of independently adjusting the electrical and thermal transport. The maximum ZT reaches 1.7 at 850 K for Ba-La-Yb triple-element-filled skutterudites, one of the best bulk TE materials at intermediate temperatures (Fig. 1).[15,17]

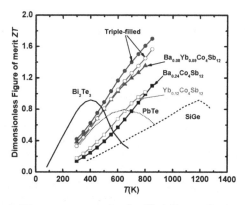

Figure 1. Figure of merit $ZT$ vs. temperature for the filled skutterudites. Data for state-of-the-practice materials are also plotted for reference.

P-type skutterudites normally require substituting Co with Fe or other transition metals with fewer valence $d$ electrons. Because of the electron deficiency caused by transition metal substitution on the Co site, filler concentration in the p-type is usually higher than that in the n-type filled CoSb$_3$. Filler concentration as well as the relative ratio of transition metals on the Co site influence the electrical and thermal transport (hence the TE performance) in these materials. The basic electronic features of the p-type have been captured in a recent theoretical study of $R$Fe$_4$Sb$_{12}$ (R=Na, K, Ca, Sr, Ba, La, Ce, Pr, Yb).[23] In $R$Fe$_4$Sb$_{12}$, both VB and CB are mainly composed of the Fe $d$ states, with one additional light Sb band near the top of VB. The $d$ bands in VB are so heavy that they pin the Fermi levels. This very feature dominates electrical transport. Electrical transport in the p-type can also be modified by the charge state and size of the fillers. For p-type Fe-Co-Ni-based skutterudites, VB and CB are also mainly composed of the $3d$ electronic states of transition metals. Some bands are split due to the different energy levels of Fe, Co, and Ni; and thus impact the band gap and the band edge density of states (DOS, $N(E)$).

## RESULTS AND DISCUSSIONS
From TE module optimization point of view, it is highly desirable to match the elastic, thermomechanical, and transport properties of the p- and n-type legs. $ZT$ values of the current p-type skutterudites have plateaued to ~ 1.0.[24-31] For the p-type skutterudites, there is still a long way to go to match the excellent thermoelectric efficiency of the n-type counterparts. One of the major shortcomings of the p-type skutterudites is their relatively low PFs. For $R$Fe$_4$Sb$_{12}$, the maximum PFs are 30~35 $\mu$W/cm-K$^2$,[27] much lower than those of the optimized n-type. In other p-type with transition metal substitution for Fe, PFs are also modest. For example, the maximum PFs are around 30 $\mu$W/cm-K$^2$ for $R_x$Fe$_3$Co$_1$Sb$_{12}$ and $R_x$Fe$_{3.5}$[Ni(Pt)]$_{0.5}$Sb$_{12}$.[28,29] In Figure 2(a), we plot the electrical transport properties of representative p- and n-type skutterudites at elevated temperatures below the onset of the bipolar effect.[14,15,17,27,28,32-34] It is noted that, comparing to the n-type, p-type skutterudites show low electrical conductivity at similar thermopower values.

11

Carrier mobility for the p-type is lower than that of the n-type, approximately by a factor of two. Thus, it is very likely that low hole mobility underpins the small PFs in p-type skutterudites.

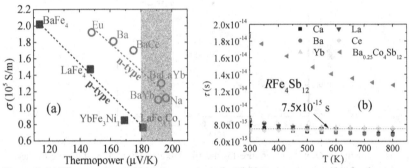

Figure 2. (a) The experimental high temperature electrical transport properties for the p-type and n-type skutterudites; (b) the electron scattering relaxation time $\tau$ for p-type $R\text{Fe}_4\text{Sb}_{12}$ skutterudites and n-type $\text{Ba}_{0.25}\text{Co}_4\text{Sb}_{12}$.

Figure 2(b) shows the carrier scattering relaxation time $\tau$ for $R\text{Fe}_4\text{Sb}_{12}$, estimated by comparing the experimental electrical conductivities with those predicted by *ab initio* $\tau$-dependent Boltzmann transport calculations.[23] Low relaxation time in the p-type skutterudites is likely caused by strong electron-phonon interactions, and hence low hole mobility in $R\text{Fe}_4\text{Sb}_{12}$. We also find the relaxation times of p-type Fe-Co-Ni-based skutterudites are within a range of $(5-8)\times10^{-15}$ s, much lower than those in the n-type. According to Ziman, under the elastic scattering and spherical energy surface approximations, the electron-phonon interactions relaxation time can be expressed as [35]

$$\tau = \frac{\hbar \rho c_l^2}{\pi \varepsilon^2 N(E_F) k_B T} = \tau_0 N(E_F)^{-1} T^{-1} \qquad (1)$$

Here $\hbar$, $\rho$, $c_l$, and $\varepsilon$ are the reduced Planck constant, mass density, sound velocity, and deformation potential, respectively. The scattering of an electron by lattice phonons depends directly on $T$ and on the electronic density of states at the Fermi level ($N(E_F)$). Parameter $\tau_0$ is approximately energy and temperature independent. As mentioned, the VB near the Fermi levels of existing Fe-Co-Ni-based p-type skutterudites are mostly composed of localized $3d$ electronic states, which are flat and heavy. This band feature causes strong electron-phonon interactions. The $3d$ transition metal substitution does not significantly alter the nature of heavy VB in skutterudites. In summary, the nature of heavy VB is a fundamental drawback of the electrical transport in $3d$ p-type skutterudites.[36]

Another disadvantage of the p-type skutterudites is the severe bipolar transport at high temperatures which is especially noticeable in the Fe-Co-Ni systems.[28,37] The bipolar conduction originated from the intrinsic excitation of electron-hole pairs, not only lowers the magnitude of thermopower due to its opposite signs for electron and hole, but also significantly increases the thermal conductivity through an electron-hole coupling which carries additional heat energy beyond phonons and charge carriers.[38,39]  These two concurrent effects apparently deteriorate the high temperature $ZT$. For instance, the bipolar thermal conductivity of $Yb_xFe_3NiSb_{12}$ is estimated > 1.5 W/m-K and constitutes 30-50% of the total thermal conductivity at 800 K. If the bipolar transport were totally suppressed, $ZT$ values could be improved by more than 20% and 50% for the Fe-Co and Fe-Ni systems, respectively.[40]  Based on a single parabolic band approximation and assuming that the acoustic phonon scatterings dominate the charge carrier scatterings, the bipolar thermal conductivity $\kappa_b$ and thermopower $\alpha$ can be written as [41]

$$\kappa_b = \frac{\sigma_n \sigma_p}{\sigma_n + \sigma_p} (\frac{k_B}{e})^2 \left[ \frac{E_g}{k_B T} + 4 \right]^2 T \quad , \quad (2)$$

$$\alpha = \frac{\alpha_n \sigma_n + \alpha_p \sigma_p}{\sigma_n + \sigma_p} \quad , \quad (3)$$

where $\sigma_n$, $\sigma_p$, $\alpha_n$, and $\alpha_p$ are the electrical conductivity and thermopower of electrons and holes; $E_g$ is the band gap, and $e$ is the fundamental charge of electrons. It is obvious that the magnitudes of $\kappa_b$ and $\alpha$ are closely related to characteristics of both the majority and the minority carriers, including the band structures, scattering processes, and so on. Eqs. (2) and (3) are also expected to be valid for heavily doped semiconductors.

Since the issues of p-type skutterudites are mostly band structure-related, we expect that altering the determining band features would lead to $ZT$ improvement. In order to significantly improve the electrical transport properties of p-type skutterudites, drastic changes in the VB are required to lower $N(E_F)$ (Eq. (1)) and therefore to achieve higher hole mobility. It has been reported that the skutterudites containing $4d$ and $5d$ transition metals,[42] such as Ru or Os, possess stronger transition metal-Sb bonds and have more dispersive bands in the VB. Following this idea, $4d$ or $5d$ transition metal-containing skutterudites were studied.[36]  Band structures near the Fermi level of $LaOs_4Sb_{12}$ are presented in Fig. 3(a). Since the $4d$ and $5d$ electronic states are spatially less localized than their $3d$ counterparts, they are expected to have stronger chemical bonding with the electronic states of the surrounding Sb atoms, which moves the $4d$ and $5d$ electronic DOS to lower energy. Consequently, there are fewer bands at the top of VB, as shown in Fig. 3(a), and therefore low $N(E)$ is expected. Stronger TM-Sb interactions also tend to make the $d$-dominated bands more dispersive. For instance, the effective masses of the top $d$-dominated band at $\Gamma$ point for $LaFe_4Sb_{12}$, $LaRu_4Sb_{12}$, and $LaOs_4Sb_{12}$ are $3.1m_0$, $1.5m_0$, and $0.9m_0$, respectively; where $m_0$ is the free electron mass.  Therefore, Ru- and Os-based skutterudites should be considered as light VB skutterudites, in contrast to the $3d$ heavy VB skutterudites. According to our Boltzmann transport calculations, however, thermopowers of $LaOs_4Sb_{12}$ and $LaRu_4Sb_{12}$ are too low (< 50 $\mu$V/K) to be useful. This implies these light VB skutterudites are not good thermoelectrics at high carrier concentrations (such as one hole per unit cell in $LaOs_4Sb_{12}$).

13

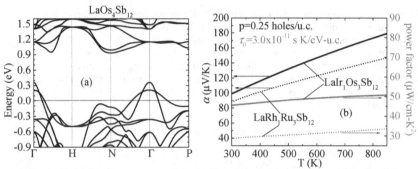

Figure 3. (a) Band structures of $LaOs_4Sb_{12}$; (b) theoretical temperature dependence of thermopower and power factor for $LaIr_1Os_3Sb_{12}$ and $LaRh_1Ru_3Sb_{12}$.

The rigid band approximation suggests, nevertheless, that high thermopower (~ 180 μV/K) is achievable in $LaOs_4Sb_{12}$ when the hole concentration can be reduced to 0.12 holes per unit cell. At this doping level, due to the light VB, $N(E_F)$ of $LaOs_4Sb_{12}$ is even lower than those of n-type skutterudites. This may lead to low electron-phonon interactions and high carrier mobility. The crux of light VB skutterudites optimization points to obtaining low hole concentration in benefit of high thermopower as well as high mobility. Experimentally this could be achieved by transition metal substitution in the p-type. Several compositions such as $LaIr_1Os_3Sb_{12}$ and $LaRh_1Ru_3Sb_{12}$ have been studied.[36] For example, $LaIr_1Os_3Sb_{12}$ is a semiconductor, and low hole concentration could be achieved through La deficiency. Ir substitution in $LaOs_4Sb_{12}$ moves the Os-bands a bit downwards and modestly increases $N(E)$; yet still results in a light VB skutterudite. Band gap of $LaIr_1Os_3Sb_{12}$ is almost the same as that of $LaOs_4Sb_{12}$, which is beneficial for mitigating the bipolar effects.

For $LaIr_1Os_3Sb_{12}$ at 850 K, the corresponding hole concentration for obtaining $\alpha$ = 180 μV/K (suggested by Ioffe as optimal for high efficiency TE materials) [43] is 0.25 holes per unit cell, under the rigid band approximation. This hole concentration is different from that of $LaOs_4Sb_{12}$ (0.12 holes per unit cell) since the VB has been altered, but still much lower than those of the n-type $CoSb_3$ skutterudites (0.4-0.6 holes per unit cell) or the $3d$ heavy VB p-type skutterudites (0.6-0.8 holes per unit cell). The estimated PFs for $LaIr_1Os_3Sb_{12}$ are over 50 μW/cm-$K^2$ at high temperatures, comparable to those of the optimized n-type $CoSb_3$ skutterudites. These high PFs come from the sufficiently high thermopower, and good electrical conductivity attributed to large relaxation time and low $N(E_F)$ in this compound. We note that the calculated PFs are not necessarily in exact agreement with the experimental data considering the assumptions used. The main idea presented here lies in the use of $4d$ and/or $5d$ transition metal to form light VB p-type skutterudites, which show promise for significant electron-phonon interactions reduction and hence the electrical conductivity improvement. It is also worth noting that Fe could be introduced into the light VB skutterudites, and the combination of both $3d$ and $4d/5d$ electronic states makes the VB $N(E)$ adjustable to obtain the optimal hole concentrations. Experiments on the light VB skutterudites are in progress.

The second strategy to improve TE properties of p-type materials is the suppression of bipolar transport at high temperatures, through increasing thermopower and decreasing the bipolar thermal conductivity. In degenerate p-type skutterudites, the hole concentration and conductivity are generally much higher than those of the electrons. Therefore Eqs. (2) and (3) can be simplified as

$$\alpha = \frac{\sigma_n}{\sigma_p}\alpha_n + \alpha_p \qquad , \qquad (4)$$

$$\kappa_b = \sigma_n(\frac{k_B}{e})^2 \left[\frac{E_g}{k_B T} + 4\right]^2 T \qquad . \qquad (5)$$

It is clear that the magnitudes of $\kappa_b$ and $\alpha$ are roughly proportional to the electron conductivity which is determined by the band structures and scattering processes. For a given hole concentration and a given VB effective mass, the influences of other band structures related parameters, including the CB effective mass and the band gap, on the bipolar thermal conductivity are analyzed. Fig. 4(a) shows $\kappa_b$ as functions of band gap and temperature, while the CB and VB effective masses are respectively set as $2.0m_0$ and $5.0m_0$ and hole concentration as $4.9\times10^{20}$ cm$^{-3}$. It is obvious that $\kappa_b$ strongly depends on both temperature and band gap. When the band gap is larger than 0.3 eV, $\kappa_b$ is smaller than 0.5 W/m-K; nevertheless, it increases rapidly with decreasing $E_g$ for $E_g$ < 0.3 eV. For samples with band gaps of 0.2 eV and 0.1 eV, typical for thermoelectric materials such as the Fe-Ni-based skutterudites, Bi$_2$Te$_3$-based materials, and so on, $\kappa_b$ almost dominates the thermal transport at high temperatures. Furthermore, the onset temperature of bipolar thermal transport also decreases with decreasing band gap. For a given band gap of 0.2 eV and a VB effective mass of 5.0 $m_0$, $\kappa_b$ as functions of temperature and CB effective mass $m_n$ is shown in Fig. 4(b). The $\kappa_b$ shows similar values at 300 K but different temperature dependences for different $m_n$. It is clear that $\kappa_b$ increases with increasing temperature and the CB effective mass, and these trends are strongly coupled to the electron (minority carrier) concentration. In particular, the density of state $N(E)$ at the band edge for minority carrier is a key factor to the bipolar thermal conduction. The effective mass of the majority carrier also exerts its influence on the bipolar effect. Usually large effective mass and high concentration of majority carrier would suppress the bipolar contribution at high temperatures.

Based these analyses, one of the keys to mitigate the bipolar transport in p-type skutterudites is the band structure modification, through increasing the band gap and/or decreasing the CB effective mass, as indicated in Eqs. (4) and (5). In fact, this remains to be very challenging due to the complexity and inter-correlations between the VB and the CB. Proper chemical substitution or doping on Sb/transition metal sites could possibly modify band structures and thereby suppress the bipolar conduction. The second possible option to minimize bipolar transport in p-type skutterudites is to reduce electron mobility. Microstructure refinement or interface modification could perhaps introduce significant scatterings of electrons (the minority carriers) but negligible influence on the transport of holes (the majority carrier). It's been demonstrated that Bi$_2$Te$_3$-based nano-composites whose onset temperature of bipolar conduction has been noticeably shifted to higher temperatures, as compared to the conventional zone melted counterpart of similar carrier concentrations.[44,45] The strategies of minimizing the bipolar transport should work synergistically with the PFs enhancement approaches in order to further improve TE performances of the p-type.

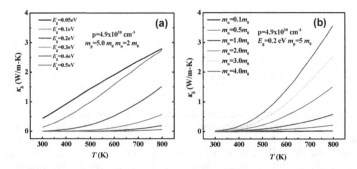

Figure 4. Estimated bipolar thermal conductivity $\kappa_b$ with respect to (a) band gap $E_g$ and (b) CB effective mass $m_n$ at a given VB effective mass and a given hole concentration.

## CONCLUSIONS

In summary, we have investigated the two major limitations in improving the TE performance of p-type skutterudites. Both of them, i.e., low power factor and severe bipolar effect, could be attributed to essential electronic band features. Power factors of the existing p-type with $3d$ electronic states are much lower than those of the optimized n-type, due to the heavy VB and low hole mobility. We propose the use of 4d/5d transition metals to form light VB skutterudites. More importantly, due to the light VB nature, weak electron-phonon interactions and high mobility could be expected in the p-type with $4d/5d$ electrons. The power factors for LaIr$_1$Os$_3$Sb$_{12}$ are estimated to be over 50 $\mu$W/cm-K$^2$ at high temperatures, higher than those of the $3d$ p-type skutterudites. In addition, the bipolar transport at elevated temperatures significantly deteriorates thermoelectric properties of the p-type, in a manner of decreasing thermopower and increasing thermal conductivity. The key to suppress the bipolar transport is to decrease conductivity of the minority carriers (electrons in the p-type), increase the band gap, decrease the CB effective mass, or decrease the electron mobility. Our work calls for vigorous experimental and theoretical work in p-type skutterudites in order to clearly discern the underlying physical mechanisms that hinder the improvement of their thermoelectric properties.

## ACKNOWLEDGEMENTS

Jihui Yang would like to acknowledge funding from the University of Washington and Kyocera Foundation. This work is also in part supported by National Basic Research Program of China (No. 2007CB607500), National Natural Science Foundation of China (Grant Nos. 11179013, 1100421, 50825205), and by the CAS/SAFEA International Partnership Program for Creative Research Teams.

# REFERENCES

1. T. M. Tritt and M. A. Subramanian, MRS Bulletin 31, 188 (2006).
2. C. Uher, Semiconductors and semimetals, in Recent Trends in Thermoelectric Materials Research I, Vol 69, edited by T. M. Tritt (Acamedic, San Diego, 2001), pp. 139-253.
3. X. Shi, W. Zhang, L. Chen, and J. Yang, Physical Review Letter 95, 185503 (2005).
4. G. Nolas, G. Slack, D. Morelli, T. Tritt, and A. Ehrlich, Journal of Applied Physics 79, 4002 (1996).
5. B. Sales, D. Mandrus, B. Chakoumakos, V. Keppens, and J. Thompson, Physical Review B 56, 15081 (1997).
6. G. Nolas, J. Cohn, and G. Slack, Physical Review B 58, 164 (1998).
7. G. Nolas, M. Kaeser, R. Littleton, and T. Tritt, Applied Physics Letters 77, 1855 (2000).
8. G. Lamberton, S. Bhattacharya, R. Littleton, M. Kaeser, R. Tedstrom, T. Tritt, J. Yang, and G. Nolas, Applied Physics Letters 80, 598 (2002).
9. M. Puyet, B. Lenoir, A. Dauscher, M. Dehmas, C. Stiewe, and E. Müller, Journal of Applied Physics 95, 4852 (2004).
10. Y. Z. Pei, L. D. Chen, W. Zhang, X. Shi, S. Q. Bai, X. Y. Zhao, Z. G. Mei, and X. Y. Li, Applied Physics Letters 89, 221107 (2006).
11. Z. Mei, J. Yang, Y. Pei, W. Zhang, L. Chen, and J. Yang, Physical Review B 77, 045202 (2008).
12. L. Xi, J. Yang, W. Zhang, L. Chen, and J. Yang, Journal of the American Chemical Society 131, 5560 (2009).
13. Jiong Yang, L. Xi, W. Zhang, L. Chen, and Jihui Yang, J. Electronic Materials 38, 1397 (2009).
14. X. Shi, H. Kong, C.-P. Li, C. Uher, J. Yang, J. R. Salvador, H. Wang, L. Chen, and W. Zhang, Applied Physics Letters 92, 182101 (2008).
15. S. Q. Bai, Y. Z. Pei, L. D. Chen, W. Q. Zhang, X. Y. Zhao, and J. Yang, Acta Materialia 57, 3135 (2009).
16. J. R. Salvador, J. Yang, H. Wang, and X. Shi, Journal of Applied Physics 107, 043705 (2010).
17. X. Shi, J. Yang, J. R. Salvador, M. F. Chi, J. Y. Cho, H. Wang, S. Q. Bai, J. H. Yang, W. Q. Zhang, and L. D. Chen, Journal of the American Chemical Society 133, 7837 (2011).
18. B. C. Sales, D. Mandrus, and R. K. Williams, Science 272, 1325 (1996).
19. V. D. Mandrus, B. Sales, B. Chakoumakos, P. Dai, R. Coldea, M. Maple, D. Gajewski, E. Freeman, and S. Bennington, Nature 395, 876 (1998).
20. R. P. Hermann, R. Jin, W. Schweika, F. Grandjean, D. Mandrus, B. C. Sales, and G. J. Long, Physical Review Letters 90, 135505 (2003).
21. J. Yang, W. Zhang, S. Q. Bai, Z. Mei, and L. D. Chen, Applied Physics Letters 90, 192111 (2007).
22. D. G. Cahill, S. K. Watson, and R. O. Pohl, Physical Review B 46, 6131 (1992).
23. J. Yang, P. Qiu, R. Liu, L. Xi, S. Zheng, W. Zhang, L. Chen, D. Singh, and J. Yang, Physical Review B 84, 235205 (2011).
24. B. Chen, J. H. Xu, C. Uher, D. T. Morelli, G. P. Meisner, J. P. Fleurial, T. Caillat, and A. Borshchevsky, Physical Review B 55, 1476 (1997).
25. X. Tang, Q. Zhang, L. Chen, T. Goto, and T. Hirai, Journal of Applied Physics 97, 093712 (2005).
26. Q. M. Lu, J. X. Zhang, X. Zhang, Y. Q. Liu, D. M. Liu, and M. L. Zhou, J. Appl. Phys. 98,

106107 (2005).
27. P. F. Qiu, J. Yang, R. H. Liu, X. Shi, X. Y. Huang, G.J. Snyder, W. Zhang, and L. D. Chen, Journal of Applied Physics 109, 063713 (2011).
28. R. Liu, J. Yang, X. Chen, X. Shi, L. Chen, and C. Uher, Intermetallics 19, 1747 (2011).
29. P. F. Qiu, R. H. Liu, J. Yang, X. Shi, X. Y. Huang, W. Zhang, L. D. Chen, Jihui Yang, and D. J. Singh, Journal of Applied Physics 111, 023705 (2012).
30. G. Rogl, A. Grytsiv, P. Rogl, E. Bauer, M. Kerber, M. Zehetbauer, and S. Puchegger, Intermetallics 18, 2435 (2010).
31. G. Rogl, A. Grytsiv, P. Rogl, E. Bauer, and M. Zehetbauer, Intermetallics 19, 546 (2011).
32. Y. Z. Pei, S. Q. Bai, X. Y. Zhao, W. Zhang, and L. D. Chen, Solid State Sciences 10, 1422 (2008).
33. L. D. Chen, T. Kawahara, X. F. Tang, T. Goto, T. Hirai, J. S. Dyck, W. Chen, and C. Uher, Journal of Applied Physics 90, 1864 (2001).
34. Y. Z. Pei, Jiong Yang, L. D. Chen, W. Zhang, J. R. Salvador, and Jihui Yang, Applied Physics Letters 95, 042101 (2009).
35. J. M. Ziman, Electrons and Phonons, (Clarendon Press, Oxford, 1960), p. 433.
36. Jiong Yang, R. Liu, Z. Chen, L. Xi, Jihui Yang, W. Zhang, and L. Chen, Applied Physics Letters 101, 022101 (2012).
37. R. Liu, P. Qiu, X. Chen, X. Huang, and L. Chen, Journal of Material Research 26, 1814 (2011).
38. P. Price, Philosophical Magazine 46, 1252 (1955).
39. H. Goldsmid, Proceedings of the Physical Society. Section B 69, 203 (2002).
40. S. Wang et al., unpublished.
41. H. Goldsmid, Thermoelectric refrigeration, Plenum press, Chap. 2, p38.
42. D. J. Singh, Semiconductors and semimetals, in Recent Trends in Thermoelectric Materials Research II, Vol 70, edited by T. M. Tritt (Acamedic, San Diego, 2001), pp. 125-177.
43. A. F. Ioffe, Semiconductor Thermoelements and Thermoelectric Cooling, (Infosearch Limited Press, London, 1957), p. 49.
44. B. Poudel, Q. Hao, Y. Ma, Y. Lan, A. Minnich, B. Yu, X. Yan, D. Wang, A. Muto, and D. Vashaee, Science 320, 634 (2008).
45. S. Wang, W. Xie, H. Li, and X. Tang, Intermetallics 19, 1024 (2011).

Mater. Res. Soc. Symp. Proc. Vol. 1490 © 2013 Materials Research Society
DOI: 10.1557/opl.2013.23

# Influence of Sn on the structural and thermoelectric properties of the type-I clathrates Ba$_8$Cu$_5$Si$_6$Ge$_{35-x}$Sn$_x$ (0 ≤ x ≤ 0.6)

X. Yan[1], E. Bauer[1], P. Rogl[2], and S. Paschen[1]
[1]Institute of Solid State Physics, Vienna University of Technology, Wiedner Hauptstr. 8–10, 1040 Vienna, Austria.
[2]Institute of Physical Chemistry, University of Vienna, Währingerstr. 42, 1090 Vienna, Austria.

## ABSTRACT

On the search for cost-competitive thermoelectric clathrates we have investigated the influence of Sn substitutions for Ge on the structural and thermoelectric properties of the type-I clathrate Ba$_8$Cu$_5$Si$_6$Ge$_{35}$. The solid solubility of Sn was found to be limited to 0.6 atoms per unit cell. A series of compounds with the nominal compositions Ba$_8$Cu$_5$Si$_6$Ge$_{35-x}$Sn$_x$ (x = 0.2, 0.4, 0.6) was synthesized in a high-frequency furnace. The samples were annealed, and subsequently ball milled and hot pressed. The hot pressed samples were characterized by X-ray powder diffraction, energy-dispersive X-ray spectroscopy and transport property measurements. Our results show that the substitution of Ge by Sn introduces vacancies at the 6d site of the type-I clathrate structure and shifts the highest dimensionless thermoelectric figure of merit ZT from 570 °C for the Sn free sample to lower temperatures. The highest figure of merit ZT = 0.42 is reached at about 320 °C for the Sn-substituted sample Ba$_8$Cu$_5$Si$_6$Ge$_{35}$Sn$_{0.6}$.

## INTRODUCTION

Intermetallic type-I clathrates have been investigated extensively in recent years as promising thermoelectric materials [1,2]. Their superior thermoelectric performance is closely related to the crystal structure, which is comprised of two kinds of polyhedral cages and guest atoms encapsulated in them [3]. The weakly bonded guest atoms can rattle in the over-sized cages, resulting in intrinsic low lattice thermal conductivities due to strongly scattered heat-carrying phonons [1,2,4-6], reduced group velocities of acoustic phonons [7], and/or a phonon filtering effect [8]. Low thermal conductivity is a prerequisite for a high dimensionless thermoelectric figure of merit, which is defined by $ZT = TS^2/(\rho\kappa)$, where $T$, $S$, $\rho$, and $\kappa$ are the absolute temperature, Seebeck coefficient, electrical resistivity, and thermal conductivity, respectively. Charge transport in clathrates is governed by the covalently bonded framework of the host atoms (cage atoms), giving rise to relatively high charge carrier mobilities. Binary type-I clathrates have a general formula A$_8$E$_{46}$, with A = alkali, alkaline, or rare-earth element, and E = group IV element. The stability of the crystal structure can be understood with the Zintl concept [9]: the valence electrons of the guest atoms A are transferred to the framework atoms E to fulfill their octet electronic configurations. Therefore, without vacancies in the structure, binary type-I clathrates could not be synthesized under normal pressure [10,11]. It has been found that the framework atoms E can be substituted by a wide variety of elements [12], e.g., transition metal and group III elements. The substitution stabilizes the crystal structure, making it possible to synthesize clathrates in an arc-melting furnace. In addition, it provides the opportunity to tune the charge carrier concentration in the system.

On the search for cost-competitive representatives, we have previously investigated the thermoelectric properties of the type-I clathrates $Ba_8Cu_5Si_xGe_{41-x}$ ($0 \leq x \leq 41$) [13]. We found that the thermoelectric properties are sensitive to details of the crystal structure that change with x in this iso-valent element substitution series. We found that the Ge-richer samples have higher ZT values. In order to further improve the thermoelectric performance of these quaternary type-I clathrates, we have now selected one composition $Ba_8Cu_5Si_6Ge_{35}$, which has a ZT value of 0.48 at about 570 °C [13], and attempted to replace Ge by Sn. Since Sn is heavier than Ge, a lower thermal conductivity is expected. In addition, the substitution is likely to create vacancies in the framework, which have been suggested to influence the thermoelectric properties [12,14-18].

## EXPERIMENT

### Sample preparation and characterization

To determine the solid solubility limit of Sn in $Ba_8Cu_5Si_6Ge_{35-x}Sn_x$, a compound with the nominal composition $Ba_8Cu_5Si_6Ge_{34}Sn_1$ was prepared in a high-frequency induction furnace from high-purity elements ($\geq 99.99$ wt.%). The sample was then annealed at 800 °C for 15 days. The annealed sample was investigated by X-ray powder diffraction (XPD) and energy-dispersive X-ray spectroscopy (EDX) to identify the phase constitution and to measure the phase composition. More details on the techniques of XPD and EDX are given in Ref. [19]. The results show that the solid solubility of Sn in $Ba_8Cu_5Si_6Ge_{35-x}Sn_x$ is limited to 0.6 atoms per unit cell.

Based on these results, a series of compounds with the nominal compositions $Ba_8Cu_5Si_6Ge_{35-x}Sn_x$ ($x = 0.2, 0.4, 0.6$), denoted below by Snx, was prepared in the induction furnace. The samples were then annealed (800 °C, 15 days) and subsequently ball milled and hot pressed (T = 800 °C, P = 56 MPa, time = 2 h). The hot pressed samples were again checked by XPD and EDX measurements. The compositions measured by EDX were normalized to 8 Ba atoms per unit cell and are listed in table I. The dependence of the physical properties on the actual Sn content $x_{Sn}$ will be discussed below.

### Physical properties

The electrical resistivity and Seebeck coefficient were measured with a ZEM-3 (ULVAC-Riko, Japan). The thermal conductivity was calculated from the thermal diffusivity $D_t$, measured by a laser flash method with a Flashline-3000 (ANTER, USA), specific heat $C_p$, and material density D using the relation $\kappa = D_t C_p D$. $C_p$ was derived from the Flashline-3000 using National Institute of Standards and Technology (NIST) steel as reference in a comparative procedure. Hall effect measurements were performed with a standard 4-point ac technique in a physical property measurement system (PPMS, Quantum Design) in the temperature range of 200-350 K in magnetic fields up to 9 T.

### DISCUSSION

### Structural properties

To disclose the influence of Sn on the thermoelectric properties of Ba-Cu-Si-Ge clathrates, the chemical and thermoelectric data of $Ba_8Cu_5Si_6Ge_{35}$ are used here as reference for our discussion. This sample is denoted as Sn0.0. Figure 1 shows the XPD spectra of $Ba_8Cu_5Si_6Ge_{35-x}Sn_x$ ($0 \leq x \leq 0.6$). The main phase can be indexed as the type-I clathrate phase. Only a small amount of diamond phase $Ge_xSi_{1-x}$, formed probably during the ball milling and hot pressing processes, appears in all samples. The phase constitution in each sample was confirmed by EDX measurements.

The lattice parameter of the clathrate phase increases slightly with increasing $x_{Sn}$ (table I), consistent with the difference between the atomic radii of Ge and Sn (figure 1, inset).

**Table I.** Sample code with the nominal Sn content, actual Sn content $x_{Sn}$, phase composition determined from EDX, lattice parameter a, charge carrier concentration $n_H$, effective mass $m^*$ and Hall mobility $\mu_H$, all derived from Hall effect measurements at 300 K. The standard deviation of the phase composition is 0.04 atoms per unit cell.

| Sample code | $x_{Sn}$ | Compositions | a (nm) | $n_H(300K)$ $(10^{26}/m^3)$ | $m^*(300K)$ $(m^*/m_e)$ | $\mu_H(300K)$ $(cm^2/Vs)$ |
|---|---|---|---|---|---|---|
| Sn0.0 | 0.0 | $Ba_{8.00}Cu_{5.03}Si_{6.00}Ge_{34.92}$ | 1.0640(1) | -8.67 | 2.26 | 7.05 |
| Sn0.2 | 0.18 | $Ba_{8.00}Cu_{4.98}Si_{5.88}Ge_{34.76}Sn_{0.18}$ | 1.0656(2) | -1.15 | 1.54 | 13.89 |
| Sn0.4 | 0.41 | $Ba_{8.00}Cu_{4.92}Si_{5.92}Ge_{34.57}Sn_{0.41}$ | 1.0664(1) | -0.96 | 1.38 | 14.11 |
| Sn0.6 | 0.55 | $Ba_{8.00}Cu_{5.01}Si_{5.93}Ge_{34.28}Sn_{0.55}$ | 1.0667(2) | -1.30 | 1.63 | 11.36 |

Rietveld refinements of the XPD data were performed with an initial model derived from single crystal X-ray diffraction data of $Ba_8Cu_5Si_6Ge_{35}$ [20]: Ba atoms fully occupy the 2a and 6c sites, Cu atoms sit at the 6d site, and the remaining sites are shared statistically by Si and Ge atoms. The results of these refinements also showed that practically no vacancies exists in the crystal structure. In the present case, since Sn is the heaviest of the framework atoms, one would expect that the location of Sn can be easily determined by XPD. However, due to the limited solid solubility, it is still not unambiguous. For the Sn0.6 sample, which has the highest Sn content, attempts were made with different structural models (see table II). We assumed that the Cu atoms exclusively occupy the 6d site with an occupation of 5 at./u.c., a value close to that measured by EDX (see table I). Then we located the Sn atoms at the framework sites 6d, 16i and 24k. The amount of Sn was fixed to the measured value of 0.55 at./u.c. Refinements were then performed with the occupations of Si, Ge, and/or vacancies (denoted by □), as well as the atomic displacement parameters (ADPs, $B_{iso}$) as free parameters. Of course, the phase composition of the clathrate phase obtained from the refinements should be close to that measured by EDX. The refinements in all cases show the same residual reliability factors. However, as seen in table II, if we located Sn at the 6d site (the top two lines in the table), the $B_{iso}$ parameter at the 2a site is too small while the one at the 6d site is too large compared to those at the other framework sites, indicating that the Sn atoms are unlikely situated at this site. If we located Sn at the 16i (middle two lines) or 24k site (bottom two lines), the results of the refinements show that when vacancies appear at the 6d site (the 4th and 5th lines in table II), the refined Si content is close to the value of 6.0 measured by EDX (table I), and the $B_{iso}$ parameters at all sites are comparable (except for the one of the rattling site 6c, see table II). The resulting amount of vacancies is about 0.3 at./u.c. in both cases. Thus, Sn appears to be situated at the 16i or 24k site, and creates vacancies at the 6d site. This finding differs from the results of the type-I clathrates $Ba_8Zn_xGe_{46-x-y}Sn_y$ ($x \approx 7.6$

21

at./u.c.) [21], where the Sn atoms either occupy the 6d site or the 24k site, and no vacancies exist in the crystal structure.

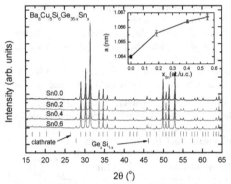

**Figure 1**. XPD patterns of $Ba_8Cu_5Si_6Ge_{35-x}Sn_x$ ($0 \leq x \leq 0.6$). The data were normalized to the highest peak (3 2 1) of the clathrate phase. The Bragg positions of the clathrate phase and diamond phase ($Ge_xSi_{1-x}$) are indicated by vertical lines. Insert: lattice parameter a of the clathrate phase vs Sn content derived from EDX, $x_{Sn}$.

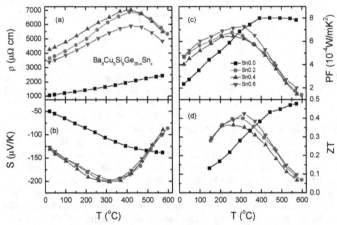

**Figure 2**. Temperature dependence of the thermoelectric properties of $Ba_8Cu_5Si_6Ge_{35-x}Sn_x$ ($0 \leq x \leq 0.6$). (a) the electrical resistivity $\rho$; (b) the Seebeck coefficient S; (c) the power factor $S^2/\rho$; and (d) the dimensionless figure of merit ZT.

**Transport properties**

22

The temperature dependent electrical resistivity $\rho(T)$ and Seebeck coefficient $S(T)$ are presented in figure 2 (a) and (b). All samples exhibit metal-like behavior, as evidenced by $d\rho/dT > 0$, at least below 400 °C. The negative sign of $S(T)$ indicates that electrons are the dominating charge carriers in all samples. Extrema are observed in both $\rho(T)$ and $S(T)$ curves of the Sn-substituted samples over the temperature range investigated, indicating that the substitution reduces the band gap $E_g$. We estimate $E_g$ from the maximum Seebeck coefficient $S_{max}$ and the temperature at which it occurs, $T_{max}$ [22] as $E_g = 2eT_{max}S_{max}$. $E_g$ is about 0.1 eV, almost independent of the Sn content in the Sn-substituted samples. The Sn substitution moves the maxima of power factors (PFs) to lower temperatures (see figure 2 (b)) due to the narrowed band gaps and high Seebeck coefficients.

Table II. Results of Rietveld refinements for XPD data of $Ba_8Cu_5Si_6Ge_{34.4}Sn_{0.6}$ with six different structural models (see text). The residual values of $R_F$ and $R_I$ are in all cases 6.9% and 7.1%, respectively. The standard deviations for the isothermal temperature factors are about $0.04\times10^2$ $nm^2$, the errors of the occupations are 0.08 at./u.c. at each sites.

| Sn at | $B_{iso}$ ($10^2$(nm$^2$) | | | | | Refined Si (at./u.c.) | Occ. at 6d site, incl. 5Cu (at./u.c.) |
|---|---|---|---|---|---|---|---|
| | Guest sites | | Framework sites | | | | |
| | 2a | 6c | 6d | 16i | 24k | | |
| 6d | 0.18 | 2.50 | 1.21 | 0.43 | 0.50 | 6.71 | 0.55Sn+0.45Si |
| | 0.25 | 2.54 | 0.84 | 0.46 | 0.49 | 5.61 | 0.55Sn+0.45□ |
| 16i | 0.34 | 2.62 | 0.44 | 0.49 | 0.43 | 7.76 | 0.52Ge+0.48Si |
| | 0.35 | 2.64 | 0.43 | 0.48 | 0.41 | 6.23 | 0.70Ge+0.30□ |
| 24k | 0.35 | 2.64 | 0.46 | 0.47 | 0.42 | 6.60 | 0.72Ge+0.28□ |
| | 0.34 | 2.63 | 0.44 | 0.48 | 0.42 | 8.29 | 0.59Ge+0.31Si |

Figure 3 shows the evolution of electrical resistivity, Seebeck coefficient, charge carrier concentration, and Hall mobility measured at 300 K, on $x_{Sn}$. The charge carrier concentration was derived from the Hall effect data using a simple one-band model, $n_H = 1/eR_H$. Table I lists the values for T = 300 K. The negative sign of $n_H(300K)$ confirms electrons as dominant charge carriers in all samples. As Sn substitutes Ge, $n_H(300K)$ decreases dramatically (see figure 3 (c)). This indicates that the substitution not only reduces the band gap, but also shifts the Fermi level to the border of the conduction band. This shift does not depend on the precise Sn content, as indicated by the similar $n_H(300K)$ in all Sn-substituted samples (see table I). The changes in the electronic band structure might be due to the expansion of the unit cell, the appearance of vacancies in the crystal structure, and/or the distortion of the cages.

With the assumption of no phonon drag contributing to the Seebeck coefficient and the relation

$$S_d(T > \theta_D) = \frac{2\pi^2 k_B^2 m^*}{e\hbar^2 (3n\pi^2)^{2/3}} T \qquad (1)$$

we derived the effective masses m* from S and $\rho$ data at 300 K, which are listed in table I. The Sn-substituted samples have lower effective masses than the Sn free sample.

With the Hall coefficient $R_H$ and the electrical resistivity $\rho$ we evaluated the Hall mobility using the relation $\mu_H = R_H/\rho$. Table I lists the values for T = 300 K (also see figure 3 (d)). The Sn-substituted samples have higher Hall mobilities than the Sn free sample which may be attributed to their lower effective masses.

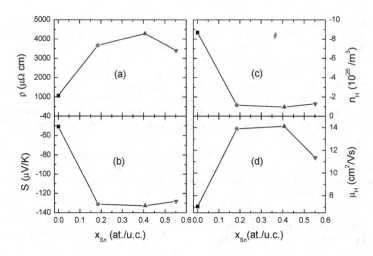

**Figure 3.** Dependence of (a) the electrical resistivity ρ, (b) the Seebeck coefficient S, (c) the charge carrier concentration $n_H$, and (d) the charge carrier mobility $\mu_H$ on the Sn content $x_{Sn}$ at T = 300 K.

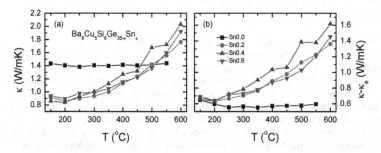

**Figure 4.** Temperature dependence of (a) the total thermal conductivity κ and (b) the difference $\kappa$-$\kappa_e$ for $Ba_8Cu_5Si_6Ge_{35-x}Sn_x$ ($0 \leq x \leq 0.6$).

Figure 4 (a) shows the temperature dependent thermal conductivity κ(T). The Sn-substituted samples have lower κ below 475 °C. However, the reduction of κ is only due to the reduced charge carrier concentration (see figure 3 (c) and table (I)). The increase of κ with increasing temperature at T > 200 °C is due to the bipolar effect [23]. By using the Wiedemann-Franz law $\kappa_e(T) = L_0 T/\rho(T)$ with $L_0 = 2.44 \times 10^{-8}$ $(V/K)^2$, we estimated the electronic contribution $\kappa_e$ from ρ. In metals and degenerate semiconductors, the lattice thermal conductivity $\kappa_{ph}$ can be calculated from $\kappa_{ph} = \kappa$-$\kappa_e$. Figure 4 (b) shows the difference $\kappa$-$\kappa_e$ versus temperature. At T > 200 °C, electrons are excited from the valence band and thus both electrons and holes participate in transports; this well-known bipolar effect enhances the Lorenz number L [23].

Therefore, the calculated $\kappa_e(T > 200\ ^\circ C)$ using $L_0$ is much smaller than the actual $\kappa_e$, i.e., the difference $\kappa$-$\kappa_e$ is much larger than the actual $\kappa_{ph}$ for the Sn-substituted samples in this temperature range. Thus, it is difficult to evaluate the influence of vacancies and distortion introduced by the Sn substitutions on the lattice thermal conductivity. However, when the conduction is in the extrinsic regime ($T < 200\ ^\circ C$), $\kappa_e$ estimated from the Wiedemann-Franz law is likely to be closer to the actual $\kappa_e$. Nevertheless, still no sizable reduction can be observed for $\kappa_{ph}$ ($\kappa_{ph} \approx \kappa$-$\kappa_e$) (see figure 4 (b)) in the Sn-substituted samples.

The dimensionless thermoelectric figure of merit ZT is potted versus temperature in figure 2 (d). The substitution of Ge by Sn shifts the maximum ZT value to lower temperatures. The highest ZT of 0.42 is reached at about 320 °C for Sn0.6. Higher ZT values are expected provided the band gap can be enlarged, e.g., by doping.

## CONCLUSIONS

The influence of Sn on the structural and thermoelectric properties of the type-I clathrate $Ba_8Cu_5Si_6Ge_{35}$ was investigated by replacing Ge with Sn. The solid solubility of Sn is limited to 0.6 at./u.c. Rietveld refinements of X-ray powder diffraction data revealed that Sn substitution creates vacancies at the 6d site, and Sn atoms occupy the 16i or 24k site in the crystal structure. Compared to the Sn free sample, the Sn-substituted samples have a higher Seebeck coefficient (double as high at 330 °C), electrical resistivity and power factor at lower temperatures (at least below 270 °C). The lattice thermal conductivity of the Sn-substituted samples is difficult to evaluate due to the bipolar effect at $T > 200\ ^\circ C$; no sizable reduction was observed at $T < 200$ °C. The Sn substitution shifts the highest ZT from 570 °C for the Sn free sample to lower temperatures. The highest ZT of 0.42 is reached at about 320 °C for the Sn-substituted sample $Ba_8Cu_5Si_6Ge_{35}Sn_{0.6}$.

## ACKNOWLEDGMENTS

The authors would like to thank M. Falmbigl and A. Grytsiv for assistance in hot pressing. We also thank M. Wass for assistance in SEM/EDX measurements. This work was supported by the Austrian Science Fund (FWF project TRP 176-N22).

## REFERENCES

1.  G. A. Slack, *CRC Handbook of Thermoelectrics*, edited by D. M. Rowe (Boca Raton, FL: CRC, 1995), pp. 407.
2.  G. A. Slack, *Mater. Res. Soc. Symp. Proc.* **47**, 478 (1997).
3.  J. Kasper, P. Hagenmuller, M. Pouchard, and C. Cros, *Science* **150**, 1713 (1965)
4.  J. L. Cohn, G. S. Nolas, V. Fessatidis, T. H. Metcalf, and G. A. Slack, *Phys. Rev. Lett.* **82**, 779 (1999).
5.  G. S. Nolas, T. J. R. Weakley, J. L. Cohn, and R. Sharma, *Phys. Rev. B* **61**, 3845 (2000).
6.  G. S. Nolas, B. C. Chakoumakos, B. Mahieu, G. J. Long, and T. J. R. Weakley, *Chem. Mater.* **12**, 1947 (2000).
7.  M. Christensen, A. B. Abrahamsen, N. B. Christensen, F. Juranyi, N. H. Andersen, K. Lefmann, J. Andreasson, C. R. H. Bal, and B. B. Iversen, *Nature Mater.* **7**, 811 (2008).

8.   H. Euchner, M. Mihalkovic, F. Gähler, M. R. Johnson, H. Schober, S. Rols, E. Suard, A. Bosak, S. Ohhashi, A.-P.Tsai, S. Lidin, C. P. Gomez, J. Custers, S. Paschen, and M. de Boissieui, Phys. Rev. B **83**, 144202 (2011)

9.   H. Schäfer, *Annu. Rev. Mater. Sci.* **15**, 1 (1985).

10.  W. Carrrillo-Cabrera, S. Budnyk, Y. Prots, and Y. Grin, *Z. Anorg. Allg. Chem.* **630**, 7226 (2004).

11.  H. Fukuoka, J. Kiyoto, and S. Yamanaka, *J. Solid State Chem.* **175**, 237–244 (2003).

12.  M. Christensen, S. Johnsen, and B. B. Iversen, *Dalton Trans.* **39**, 978 (2010).

13.  X. Yan, E. Bauer, P. Rogl, and S. Paschen, submitted *Phys. Rev. B.*

14.  X. Shi, J. Yang, S. Bai, J. Yang, H. Wang, M. Chi, J. R. Salvador, W. Zhang, L. Chen, and W. Wong-Ng, *Adv. Funct. Mater.* **20**, 755 (2010).

15.  S. Johnsen, A. Bentien, G. K. H. Madsen, and B. B. Iversen, *Chem. Mater.* **18**, 4633 (2006).

16.  A. Bentien, S. Johnsen, and B. B. Iversen, *Phys. Rev. B* **73**, 09403 (2006).

17.  L. T. K. Nguyen, U. Aydemir, M. Baitinger, E. Bauer, H. Borrmann, U. Burkhardt, J. Custers, A. Haghighirad, R. Höfler, K. D. Luther, F. Ritter, W. Assmus, Y. Grin, and S. Paschen, *Dalton Trans.* **39**, 1071 (2010).

18.  M. Hokazono, H. Anno, and K. Matsubara, *Mater. Trans.* **46(07)**, 1485 (2005).

19.  X. Yan, M. X. Chen, S. Laumann, E. Bauer, P. Rogl, R. Podloucky, and S. Paschen, *Phys. Rev. B* **85**, 165127 (2012).

20.  X. Yan, A. Grytsiv, G. Giester, E. Bauer, P. Rogl, and S. Paschen, *J. Electron. Mater.* **40**, 589 (2011).

21.  M. Falmbigl, A. Grytsiv, P. Rogl, X. Yan, E. Royanian, and E. Bauer, submitted to *Dalton Trans.* DOI: 10.1039/c2dt32049e.

22.  G. J. Goldsmid and J. W. Sharp, *J. Electron. Mater.* **28**, 869 (1999).

23.  P. J. Price, *Philos. Mag.* **46**, 1252 (1955).

Mater. Res. Soc. Symp. Proc. Vol. 1490 © 2013 Materials Research Society
DOI: 10.1557/opl.2013.287

# Attrition-enhanced nanocomposite synthesis of indium-filled, iron-substituted skutterudite antimonides for improved performance thermoelectrics

James Eilertsen[a, b, c], Matthias Trottmann[b], Sascha Populoh[b], Romain Berthelot[a], Charles M. Cooke[c,d], Michael K. Cinibulk[c], Simone Pokrant[b], Anke Weidenkaff[b], M. A. Subramanian[a]
[a]Department of Chemistry, Oregon State University, Corvallis, OR 97331, USA
[b]Empa,Solid State Chemistry and Catalysis, Ueberlandstrasse 129, CH-8600 Duebendorf, Switzerland
[c]Materials and Manufacturing Directorate, Air Force Research Laboratory, Wright-Patterson AFB, OH 45433-7817, USA
[d]UES, Inc., Dayton OH 45432, USA

## ABSTRACT

Nanostructuring has been the foremost approach to the manufacture of high-performance thermoelectric materials for nearly a decade. This study explores a novel nanostructuring technique, attrition-enhanced nanocomposite synthesis, in maximum indium-filled, iron-substituted cobalt antimonide skutterudites. $In_{0.3}Fe_{0.8}Co_{3.2}Sb_{12}$ was synthesized and subjected to varying degrees of mechanical attrition (via ball milling). These samples exhibited increased indium precipitation coincident with the duration of mechanical attrition. Indium readily diffused through the skutterudite crystal structure and rapidly precipitated forming 20-50 nm-sized indium-rich inclusions during sintering.

## 1. INTRODUCTION

The prospect of catastrophic social and economic consequences triggered by global climate change has stimulated unprecedented demand for a host of diverse, clean, and sustainable energy technologies.[1] This demand may be satiated in part by high-efficiency thermoelectric materials.[2] Consequently, decades of research have focused on enhancing their efficiency. Thermoelectric efficiency is typically reported as a dimensionless figure-of-merit ($zT$) at a specified temperature ($T$) (Eq. 1):

$$zT = \frac{\sigma S^2}{\kappa_T} T \qquad \text{Eq. 1}$$

where $\kappa_T = \kappa_L + \kappa_e$           Eq. 1a

The $zT$ is dependent on the electrical conductivity ($\sigma$), Seebeck coefficient ($S$), and lattice ($\kappa_L$) and electronic ($\kappa_e$) thermal conductivities of the material. Enhanced power-factors ($\sigma S^2$) are often observed in moderately to heavily doped semiconductors which can possess both optimum charge carrier concentration and charge carrier mobility while maintaining a relatively high Seebeck coefficient.[3,4]

The skutterudite crystal structure is comprised of a cage-like framework of transition metal-pnicogen, and pnicgoen-pnicogen bonds. The pnicogens form two large icosahedral voids sites per unit cell, which can be filled with a wide-array of disparate elements. These interstitials are loosely bound and reduce the lattice thermal conductivity substantially while also providing considerable charge carrier density thereby enhancing the electrical conductivity significantly.[5-14] The $zT$ of skutterudites, therefore, can be enhanced strongly by icosahedral void-site filling.

In addition, nanocomposite thermoelectrics – materials consisting of a polycrystalline matrix with nano-sized inclusions – also exhibit strongly enhanced $zT$'s.[15-20] Attrition-enhanced nanocomposite synthesis (AENS) is a novel technique used to produce nano-sized inclusions by precipitating interstitials from the filled void-sites of the skutterudite crystal structure.[21] The primary requisite of AENS is that the interstitial filler is metastable, as in the $Ga_xCo_4Sb_{12}$ and $In_xCo_4Sb_{12}$ compositions.[21-24] These compositions are then subjected to heavy attrition via mechanical milling where they are reduced to heavily deformed, fine-grained powders containing dislocations and other defects. The sample powders are sintered, and heat-treated to precipitate the metastable fillers from the icosahedral void-sites. In our previous work, we theorized that the combination of the open cage-like crystal structure, high dislocation and other defect concentrations, and minimized interstitial diffusion path lengths indeed enhance the diffusion kinetics and distribution of the precipitated interstitial – thus enabling the wide-spread dissemination of nano-sized phonon-scattering precipitates.[21] The technique, however, has not been fully optimized.

This research explores the AENS procedure with a maximally indium-filled, iron-substituted cobalt antimonide composition. The sample was divided and subjected to various durations of ball milling, and rapidly densified using plasma-assisted sintering (SPS). It was found that with the duration of ball milling: 1) the amount of precipitated indium increased, 2) the succeeding indium-rich precipitates decreased in size from approximately 50 to 20 nm, 3) the electronic carrier concentration decreased, and 4) the thermal conductivity increased coincident with a reduction of indium concentration in the icosahedral void sites of the skutterudite crystal structure.

## 2. EXPERIMENTAL

A large 20-gram batch of void-site-filled and cobalt-site-doped $In_{0.3}Fe_{0.8}Co_{3.2}Sb_{12}$ was synthesized from elemental indium (Aldrich, 100 mesh, 99.99 %), iron (Alfa Aesar, 200 mesh, 99+%), cobalt (Aldrich, < 2 um, 99.8 %), and antimony (Alfa Aesar, 100 mesh, 99.5 %) under an atmosphere of antimony vapor and a 95%:5% $N_2:H_2$ gas mixture according to the procedure developed by He et al.[25] The batch was divided into four samples and mechanically milled in a Fritzch Pulverisette 357 planetary ball mill for different durations (0, 10, 20, and 40 hours) at 400 rpm with silicon nitride bowls and grinding balls according to the procedure developed by us.[21] The milled powder was loaded into 20-mm graphite dies and sintered using an FCT System GmbH SE607 spark-plasma sintering furnace: The samples were purged with argon and heated to 600 °C under vacuum with a heating ramp rate of 400 °C/min, sintered in AC mode using $10^3$ A current with $10^4$ N of applied force, and the sample chamber was back-filled with argon and rapidly cooled to room temperature at approximately 40 °C/min. Approximately 85 – 90% of the theoretical density was achieved in the sintered pellets.

Powder X-ray diffraction (XRD) data were collected on as-synthesized, milled and sintered, and heat-treated samples using a Rigaku Ultima IV Multipurpose X-ray Diffraction System. The samples were loaded onto an oriented Si single-crystal sample holder (MTI Corporation) with nearly zero background to maximize the possibility of detecting impurity phases. Diffraction patterns were collected with a fixed-time scan rate of 0.01 °step$^{-1}$ and 0.1 sec step$^{-1}$ from 10 to 120 °2θ.

The diffraction data were analyzed using the Le Bail technique as implemented in the Fullprof program.[26,27] Peak shape was described by a Pseudo-Voigt function with additional asymmetric parameters for low-angle domain peaks (below 40 °2θ), and the background level was fitted with a linear interpolation between a set of 40 to 60 given points with refineable heights. Moreover, both fractured and polished samples (polished with submicron slurry) were

analyzed using an FEI NovaNanoSEM 230 scanning electron microscope (SEM) to determine grain structure and confirm interstitial precipitation. The bulk and interstitial compositions were determined by energy-dispersive X-ray spectroscopy (EDS).
Carrier concentration was determined from Hall mobility and electrical conductivity measurements on a lab-built setup at room temperature. Principal thermoelectric properties were measured from 300 to 650 K. Electrical conductivity and Seebeck coefficient data were collected using an Ulvac-Riko ZEM 3 under static helium atmosphere. Thermal diffusivity ($\alpha$) and specific heat ($C_p$) data were collected under flowing $N_2$ using a Netzsch LFA 457 Micro Flash, and a Mettler Toledo 821e Differential Scanning Calorimeter, respectively. Total thermal conductivity was determined from the relation $\kappa_T = c_p\,\alpha\,d$, where $d$ is the sample bulk density.

## 3. DISCUSSION

### 3.1 X-ray diffraction and microstructural data

X-ray diffraction data of as-synthesized and sintered, and 40-hour-milled and sintered samples are shown (Fig. 1). X-ray diffraction data reveal the indium-filled, iron-substituted skutterudite crystal structure was synthesized. A small Sb impurity is observed in the as-synthesized and sintered and ball-milled and sintered $In_{0.3}Fe_{0.8}Co_{3.2}Sb_{12}$ samples, and is attributed to the excess antimony vapor employed during synthesis. An $FeSb_2$ impurity is detected in the as-synthesized and sintered sample; however, it was no longer detected in the ball-milled and sintered samples. A small Sb impurity is observed in all samples.
Lattice parameter data derived from LeBail refinements of as-synthesized and sintered, milled and sintered samples are shown (Figure 3a). A significant lattice-parameter contraction is observed in the sample following sintering as observed in our previous work on the $In_{0.1}Co_4Sb_{12}$ skutterudites treated to a similar nano-structuring procedure.[21]

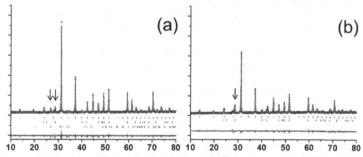

Figure 1: XRD data of the as-synthesized and sintered (a), and 40-hour ball-milled and sintered $In_{0.3}Fe_{0.8}Co_{3.2}Sb_{12}$ samples. A small $FeSb_2$ is observed in the as-synthesized and sintered sample (b) and a small Sb impurity is observed in both (black arrows).

Fractured and polished 0-, 10-, 20-, and 40-hour ball-milled samples analyzed by SEM exhibited decreasing bulk grain size with increasing ball milling time. Indium-rich precipitates were detected in the 10-, 20-, and 40-hour samples: their composition was confirmed with EDS (Fig. 2). The precipitates ranged in size from approximately 50 nm in the 10-hour sample to 20 nm in the 20- and 40-hour samples.

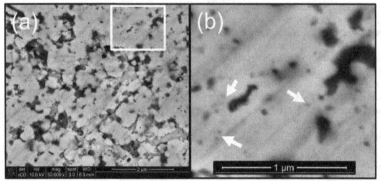

Figure 2:  SEM image of the 40-hour ball-milled and sintered $In_{0.3}Fe_{0.8}Co_{3.2}Sb_{12}$ polished sample.  The sample was replete with 20 - 30 nm inclusions (white arrows, b); their composition was determined to be indium rich from EDS.

### 3.2 Electronic property data

The electronic property data of the $In_{0.3}Fe_{0.8}Co_{3.2}Sb_{12}$ samples are shown (Fig. 3).  The charge carrier concentration drops with ball milling and is coincident with the observed lattice parameter contraction.  All compositions are degenerate; however, a clear correlation between the duration of milling and the observed drop in electrical conductivity is evident (Fig. 3b).  The magnitude of the drop generally trends with the duration of milling: however, the slight enhancement in the 40-hour sample may be due to a slight improvement in densification during spark-plasma sintering.  The Seebeck coefficient data is shown (Fig. 3c), and is observed to drop after ball milling.  A definative trend with the duration of ball milling is not observed, however.  There is little varitiaton in the Seebeck coefficients of the milled samples; the as-syntehsized and sintered sample, however, exhibits a marekdly higher value.  The combined effect of partial indium precipitaiton and charge-filtering at the nanoinclusions matrix interfaces may account for the incongruity in Seebeck, carrier concentraion and conductivity data.

### 3.3 Thermal conductivity

The total and lattice thermal conductivity data of the samples are shown (Fig. 3d).  A strong increase in thermal conductivity is observed with the duration of ball milling, coincident with a reduced concentration of indium in the icosahedral void-sites of the skutterudite crystal structure.

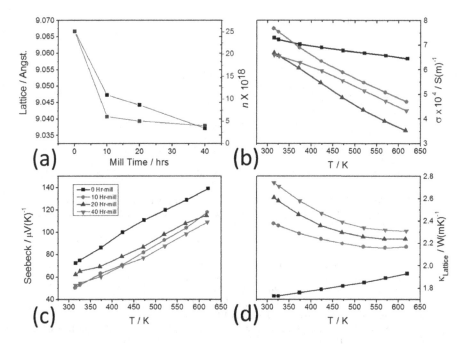

Figure 3: Lattice parameter and carrier concentration (a), electrical conductivity (b), Seebeck coefficient (c), and lattice thermal conductivity (d) of the 0-, 10-, 20-, and 40-hour ball-milled samples. All data points to a reduced indium concentration in the skutterudite crystal structure with an increase in the duration of ball milling.

## 4. CONCLUSION

Indium-filled and iron-substituted cobalt antimonde skutterudites were subjected to an attrition-enhanced nanocomposite synthesis (AENS) procedure and sintered using spark-plasma. It is evident that indium precipitates from the skutterudite crystal structure to form nano-sized indium-rich inclusions.

**Acknowledgements**
    This research was supported by a grant from the National Science Foundation (DMR 0804167), and a Marie Curie Fellowship (PCOFUND-GA-2010-267161).

(1)    Snyder, G. J.; Toberer, E. S. *Nature Materials* **2008**, *7*, 105.
(2)    Bell, L. E. *Science* **2008**, *321*, 1457.
(3)    Rowe, D. M. *CRC handbook of thermoelectrics*; CRC Press: Boca Raton, FL, 1995.

(4)     Rowe, D. M.; Bhandari, C. M. *Modern thermoelectrics*; Reston Pub. Co.: Reston, Virginia, 1983.
(5)     Nolas, G. S.; Morelli, D. T.; Tritt, T. M. *Annual Review of Materials Science* **1999**, *29*, 89.
(6)     Tritt, T. M.; Subramanian, M. A. *Mrs Bulletin* **2006**, *31*, 188.
(7)     Sales, B. C.; Mandrus, D.; Chakoumakos, B. C.; Keppens, V.; Thompson, J. R. *Physical Review B* **1997**, *56*, 15081.
(8)     Keppens, V.; Mandrus, D.; Sales, B. C.; Chakoumakos, B. C.; Dai, P.; Coldea, R.; Maple, M. B.; Gajewski, D. A.; Freeman, E. J.; Bennington, S. *Nature* **1998**, *395*, 876.
(9)     Kleinke, H. *Chemistry of Materials* **2010**, *22*, 604.
(10)    Sootsman, J. R.; Chung, D. Y.; Kanatzidis, M. G. *Angewandte Chemie-International Edition* **2009**, *48*, 8616.
(11)    Vineis, C. J.; Shakouri, A.; Majumdar, A.; Kanatzidis, M. G. *Advanced Materials* **2010**, *22*, 3970.
(12)    Nolas, G. S. In *Thermoelectric Materials 1998 - the Next Generation Materials for Small-Scale Refrigeration and Power Generation Applications*; Tritt, T. M. K. M. G. M. G. D. L. H. B., Ed. 1999; Vol. 545, p 435.
(13)    Iversen, B. B.; Palmqvist, A. E. C.; Cox, D. E.; Nolas, G. S.; Stucky, G. D.; Blake, N. P.; Metiu, H. *J ournal of Solid State Chemistry* **2000**, *149*, 455.
(14)    Sales, B. C.; Mandrus, D.; Williams, R. K. *Science* **1996**, *272*, 1325.
(15)    Hsu, K. F.; Loo, S.; Guo, F.; Chen, W.; Dyck, J. S.; Uher, C.; Hogan, T.; Polychroniadis, E. K.; Kanatzidis, M. G. *Science* **2004**, *303*, 818.
(16)    Martin, J.; Wang, L.; Chen, L.; Nolas, G. S. *Physical Review B* **2009**, *79*.
(17)    Kanatzidis, M. G. *Chemistry of Materials* **2010**, *22*, 648.
(18)    Girard, S. N.; He, J.; Li, C.; Moses, S.; Wang, G.; Uher, C.; Dravid, V. P.; Kanatzidis, M. G. *Nano Letters* **2010**, *10*, 2825.
(19)    Han, M. K.; Hoang, K.; Kong, H. J.; Pcionek, R.; Uher, C.; Paraskevopoulos, K. M.; Mahanti, S. D.; Kanatzidis, M. G. *Chemistry of Materials* **2008**, *20*, 3512.
(20)    Chen, G.; Dresselhaus, M. S.; Dresselhaus, G.; Fleurial, J. P.; Caillat, T. *International Materials Reviews* **2003**, *48*, 45.
(21)    Eilertsen, J.; Rouvimov, S.; Subramanian, M. A. *Acta Materialia* **2012**, *60*, 2178.
(22)    Eilertsen, J.; Berthelot, R.; Sleight, A. W.; Subramanian, M. A. *Journal of Solid State Chemistry* **2012**, *190*, 238.
(23)    Xiong, Z.; Xi, L. L.; Ding, J.; Chen, X. H.; Huang, X. Y.; Gu, H.; Chen, L. D.; Zhang, W. Q. *Journal of Materials Research* **2011**, *26*, 1848.
(24)    Eilertsen, J.; Li, J.; Rouvimov, S.; Subramanian, M. A. *Journal of Alloys and Compounds* **2011**, *509*, 6289.
(25)    He, T.; Chen, J.; Rosenfeld, H. D.; Subramanian, M. A. *Chemistry of Materials* **2006**, *18*, 759.
(26)    Lebail, A.; Duroy, H.; Fourquet, J. L. *Mater. Res. Bull.* **1988**, *23*, 447.
(27)    Rodríguez-Carvajal, J. *Physica B: Condensed Matter* **1993**, *192*, 55.

Mater. Res. Soc. Symp. Proc. Vol. 1490 © 2013 Materials Research Society
DOI: 10.1557/opl.2013.217

# Thermoelectric behaviour of p- and n- type Ti-Ni-Sn half Heusler alloy variants and their amorphous equivalents

Song Zhu[1], Satish Vitta[2] and Terry M. Tritt[1]
[1] Department of Physics & Astronomy
Clemson University, Clemson 29634 SC, USA.
[2] Department of Metallurgical Engineering and Materials Science
Indian Institute of Technology Bombay, Mumbai 400076, India.

## ABSTRACT

Ti-Ni-Sn type half-Heusler alloys which have the versatility to be either p- or n-type depending on the type of substitution, have been synthesized and investigated in the present work. The added advantage of doping them with multiple elements is that they will be amenable to bulk amorphous phase formation. The hole doped alloys were predominantly single phase with a cubic structure, while the electron doped alloys were found to have minor additional phases. All the alloys exhibit extremely weak metallic-like or degenerate semiconductor transport behaviour in the temperature range 20 K to 1000 K. The resistivity of p-type alloys exhibits semi-metallic-to-semiconducting transition at ~ 500 K while the n-type alloys exhibit a weak metallic-like behaviour in the complete temperature range. The Seebeck coefficient has strong temperature dependence with a maximum of 45 μV K$^{-1}$ in the temperature range 600-700 K in the p-type alloys. The n-type alloys however exhibit a linear variation of the Seebeck coefficient with temperature. The total thermal conductivity of the alloys increases with increasing temperature without any peak at low temperatures indicating significant disorder induced scattering. The p-type alloys have the lowest thermal conductivity compared to the n-type alloys. These alloys become amorphous after pulsed laser deposition except one alloy which exhibits compensated transport behaviour.

## INTRODUCTION

The single most important parameter that has been driving the development of new thermoelectric materials is the 'figure-of-merit, Z' although other material parameters such as their ability to be processed into different shapes, mechanical strength, availability of suitable contact materials are also equally important. The figure-of-merit Z defined as $\alpha^2\sigma/(\kappa_e+\kappa_l)$, where $\alpha$ is the Seebeck coefficient, $\sigma$ the electrical conductivity and $\kappa_e$ and $\kappa_l$ are electronic and lattice thermal conductivities respectively and Z depends on two competing phenomena – charge transport and heat transport which have an opposing functional dependence. Hence any materials development strategy needs to be assessed in terms of a balance between the charge carrier mobility, $\mu$ or effective mass m*of the charge carrier and the lattice thermal conductivity $\kappa_l$. The thermal conductivity can be significantly reduced by increasing boundary scattering, decreasing the effective grain size,[1,2] but this strategy can also lead to an increase in carrier scattering resulting in an insignificant change in Z. This strategy to increase Z however is predicted to be effective in a special class of materials wherein the phonon mean free path becomes significantly larger than the charge carrier mean free path. A typical case is the half-Heusler alloy, $Zr_{0.5}Hf_{0.5}NiSn$ wherein a lowering of lattice thermal conductivity is predicted due to a decrease in the grain size. An extension of this argument leads to the conclusion that this is likely to happen till the effective grain size is reduced and it equals the lattice constant, i.e. the amorphous structure limit. This is based on the assumption that band structure does not change and m* remains high. In the case of half-

Heusler alloys this has been predicted to happen at high temperatures.[3] However, a significant problem to test this hypothesis is the synthesis of amorphous half-Heusler alloys as they have a simple crystal structure which is extremely difficult to avoid during normal liquid-to-solid transformation processes. Hence, the objective of our work has been to design half-Heusler alloy compositions which will be amenable to amorphous phase formation during transformation processing. The alloy compositions have been designed based on one of the bulk amorphous phase formation criteria, viz, high viscosity liquid phase formation with low temperature dependence of viscosity.[4] This can be achieved by ensuring that the liquid will have a dense packed structure composed of widely varying atomic sizes, different elements. This criterion for half-Heusler alloys also facilitates designing chemical compositions which can be both electron and hole doped. Accordingly, 5 different half-Heusler alloy variants based on Ti-Ni-Sn which will exhibit p-type, n-type and compensated charge transport behaviours have been selected using a total of 12 different elements. The atomic radius in these alloys varies from 0.166 nm to 0.125 nm. The thermoelectric properties of these different alloys as a function of temperature have been investigated up to 900 K.

**EXPERIMENTAL METHODS**

All the 5 different alloys have been synthesized from high purity elements using the conventional arc melting process in an inert atmosphere. The alloys were then annealed in vacuum at 800 C for 120 h in order to homogenize the chemical composition. The homogenized alloy ingots were then ground to powder and consolidated into a 12.7 mm diameter highly dense pellet using a spark plasma sintering process. The phases present, structure and chemical composition of the pellets were characterized by X-ray diffraction and scanning electron microscopy. The electrical resistivity $\rho$ in the temperature range 20 K to 900 K was determined using the standard 4-probe technique. The low temperature resistivity (T < 300 K) and Seebeck coefficient were measured using a custom built system[5] at temperature intervals of 0.5 K while the high temperature data was collected using a commercial system, ZEM 2 at temperature intervals of 25 K to 50 K. The low temperature thermal conductivity was determined using a custom built system[6] whiles the high temperature thermal diffusivity was measured using a commercial system, Netzsch LFA 457. The thermal conductivity was determined using the relation $\kappa = dC_p\rho_D$ , where d is the thermal diffusivity, $C_p$ the heat capacity and $\rho_D$ the mass density of the pellet.

**RESULTS AND DISCUSSION**

The X-ray diffraction patterns from both as prepared and spark plasma sintered alloys were found to be identical and exhibit several peaks indicating that the alloys are completely crystalline. The cooling rate employed in the arc melting process is insufficient to suppress crystallization during liquid transformation into solid and results in the formation of crystalline phases. The X-ray diffraction patterns from the different alloys shown in Figure 1 have several peaks which could be identified with the cubic $F\overline{4}3m$ phase, commonly observed in half-Heusler alloys.

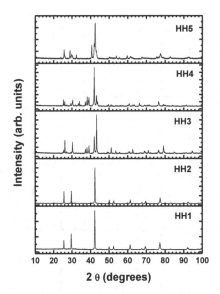

10  20  30  40  50  60  70  80  90  100
**2 θ (degrees)**

**Figure 1** The X-ray diffraction patterns from the different alloys shows crystalline peaks indicating lack of presence of any amorphous phase, in agreement with the processing conditions.

The HH1 and HH2 alloys have essentially a single cubic phase, $F\bar{4}3m$ corresponding to TiNiSn structure, while the other alloys HH3, HH4 and HH5 have other minor phases apart from the predominant cubic phase. The minor phases were also found to be cubic but with a different lattice parameter, indicating that the microstructure consists of variants of the Heusler alloy TiNiSn. These results clearly show that spark plasma sintering process does not modify the phases in as-prepared alloy but only consolidates the powders into a dense, solid pellet. The sharp peaks in the X-ray diffraction pattern indicate the presence of large grains with an average size of > 2 μm. The chemical composition of the alloys determined in the scanning electron microscope using x-ray analysis is given in Table 1 along with the expected compositions. It is found that there is no significant loss of any element in all the alloys except in HH5 alloy wherein Bi is found to be significantly lower compared to the expected value.

**Table 1**. The chemical composition of different alloys, at. %, determined by energy dispersive X-ray analysis in the scanning electron microscope is given along with the expected values.

|  | Ti | Zr | Nb | Ni | Cu | Mn | Sc | Co | Sn | In | Sb | Bi |
|---|---|---|---|---|---|---|---|---|---|---|---|---|
| Expected | 16.6 | 16.6 |  | 26.6 |  |  |  | 6.7 | 33.3 |  |  |  |
| HH1 |  |  |  |  |  |  |  |  |  |  |  |  |

| | | | | | | | | | | | | |
|---|---|---|---|---|---|---|---|---|---|---|---|---|
| Actual | 15.6 | 16.8 | | 28.8 | | | | 6.8 | 30.5 | | | |
| Expected | 16.6 | 8.3 | | 33.3 | | | 8.3 | | 31.7 | 1.7 | | |
| HH2 | | | | | | | | | | | | |
| Actual | 13.5 | 6.8 | | 31.4 | | | 9.7 | | 30.9 | 1.0 | | |
| Expected | 26.6 | | 6.7 | 26.6 | 6.7 | | | | 31.7 | | 1.7 | |
| HH3 | | | | | | | | | | | | |
| Actual | 24.3 | | 5.9 | 27.0 | 5.7 | | | | 31.6 | | 1.5 | |
| Expected | 16.6 | | 16.6 | 26.6 | 6.7 | | | | 31.7 | | 1.7 | |
| HH4 | | | | | | | | | | | | |
| Actual | 15.6 | | 12.3 | 27.7 | 5.1 | | | | 31.2 | | 1.5 | |
| Expected | 6.7 | | | 33.3 | | 26.6 | | | 30.0 | | | 3.3 |
| HH5 | | | | | | | | | | | | |
| Actual | 6.2 | | | 33.9 | | 36.4 | | | 30.6 | | | 0.4 |

The electrical resistivity of the alloys in the temperature range 20 K to 900 K is shown in Figure 2. The temperature dependence is extremely small considering that the temperature varies from 20 K to 900 K. The alloys exhibit a weak semiconducting to weak metallic behaviour typical of half-Heusler alloys. These alloys in general are known to be semiconducting with a small gap which may be direct or indirect. Substitution at different sites changes the band structure and leads to either electron or hole doped behaviour which in some cases can exhibit even a metallic like behaviour.[7,8]

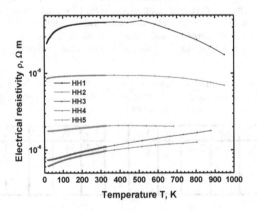

**Figure 2** The electrical resistivity of the alloys exhibits very weak temperature dependence for all the alloys. The behaviour varies from being weakly metallic to semiconducting.

The resistivity of HH1 and HH2 alloys exhibits semi-metallic to semiconducting transition at ~ 500 K while the HH3 and HH4 alloys exhibits a weak metallic like behaviour in the complete temperature range. The resistivity of all the 5 alloys is lower than the base Ti-Ni-Sn alloy which is indirect gap semiconductor with ~ $10^{-4}$ $\Omega$ m. The Seebeck coefficient $\alpha$ on the other hand has a strong temperature dependence as seen in Figure 3. The alloys HH1 and HH2 have a positive $\alpha$ above room temperature with a maximum at ~ 600 K and 700 K respectively. The alloys HH3 and HH4 exhibit a negative $\alpha$ which increases monotonically with temperature. The alloy HH5 exhibits a very unusual behaviour – the value of $\alpha$ is close to 0 in the whole temperature range with no temperature dependence, similar to a charge compensated material with no band gap.

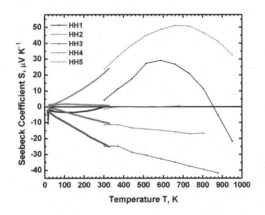

**Figure 3** The Seebeck coefficient S variation with temperature in the p-type alloys, HH1 and HH2 is strong compared to that in the n-type alloys HH3 and HH4. The HH5 alloy shows no Seebeck phenomenon indicative of a compensated behaviour.

These results can be rationalized based on the nature of substitutions in the basic Ti-Ni-Sn half-Heusler alloy. In HH1 partial substitution of Ni with Co renders it electron deficient while Zr substitution for Ti is isoelectronic in nature. Hence the alloy behaves as a hole doped system. In the case of HH2 alloy, both Sc substitution for Ti/Zr as well as In substitution for Sn renders it an electron deficient hole doped behaviour. The electron doped behaviour in the case of HH3 and HH4 is a consequence of Nb, Cu and Sb substitution for Ti, Ni and Sn respectively. The compensated behaviour observed in the case of HH5 is due to Mn substitution and Bi substitution which oppose each other and this can be seen even in the electrical resistivity which has almost no temperature dependence. The power factor, $\alpha^2\sigma$ of the alloys is shown in Figure 4 for all the alloys in the complete temperature range of investigation. The power factor of alloys HH2 and HH3, p-type and n-type, is the highest reaching a value of 200 $\mu Wm^{-1}K^{-2}$ and 900 $\mu Wm^{-1}K^{-2}$ at 700 K and 900 K respectively for the two alloys.

37

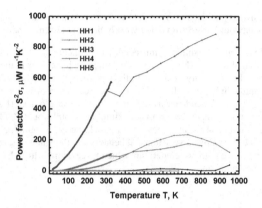

**Figure 4** The Power factor of HH2 and HH3 alloys exhibits the maximum compared to the other alloys and they show large temperature dependence.

The total thermal conductivity measured up to 300 K and that derived from the thermal diffusivity for T > 300 K is shown in Figure 5. In all the cases, the absolute value of $\kappa_{total}$ is found to be high at all temperatures, clearly showing that this parameter needs to be reduced in order to improve the figure-of-merit Z.

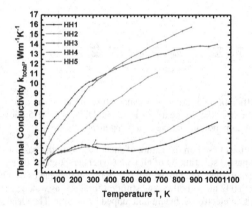

**Figure 5** The total thermal conductivity of all the alloys increases with temperature in the range 20 K to 1000 K with the lowest thermal conductivity being exhibited by the p-type alloys, HH1 and HH2.

Since it is known that the total thermal conductivity of a material should be lowest in the amorphous state,[9] the alloys have been amorphized by pulsed laser deposition into a thin film form. The structure in these alloy films has been determined by X-ray diffraction and is shown in Figure 6. It is seen that all the alloys except HH5 are completely amorphous with no discernable peaks. This indicates the inherent resistance of HH5 alloy for amorphization even when transformed into a solid at the highest possible cooling rates.

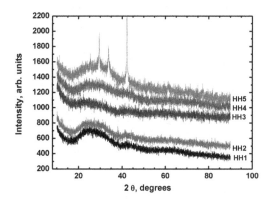

**Figure 6** The X-ray diffraction pattern from the different alloy thin films shows an amorphous character except the HH5 alloy which has crystalline peaks indicating its inherent resistance for amorphization.

The electronic and thermal transport investigation of these amorphous and crystalline thin films is currently under investigation.

## CONCLUSIONS

Half-Heusler alloys based on Ti-Ni-Sn which exhibit both p-type and n-type transport behaviour have been synthesized. The thermoelectric properties of these alloys indicate that their thermal conductivity is still high to have a high figure-of-merit. In order to reduce the thermal conductivity the alloys have been amorphized by pulsed laser deposition. These amorphous alloys should exhibit the lowest thermal conductivity and hence a high figure-of-merit at high temperatures.

## ACKNOWLEDGEMENTS

The authors wish to acknowledge the Fulbright-Nehru Foundation for providing financial assistance to SV in the form of Senior Research Fellowship. The authors (SZ and TMT) acknowledge financial support from a DOE/EPSCoR Implementation grant #DE-FG02-04ER-46139, support from the SC EPSCoR Office/Clemson University cost sharing.

## REFERENCES

1. J.W. Sharp, S.J. Poon and H.J. Goldsmid, Phys. Stat. Sol., (a) <u>187</u> 507 (2001)
2. S. Bhattacharya, T.M. Tritt, Y. Xia, V. Ponnambalam, S.J. Pon and N. Thadani, Appl. Phys. Lett., <u>81</u> 43 (2002)
3. G.S. Nolas and H.J. Goldsmid, Phys. Stat. Sol., (a) <u>194</u> 271 (2002)
4. R. Busch, J. Shroers and W.H. Wang, MRS Bulletin, <u>32</u> 620 (2007)

5. A.L. Pope, R.T. Littleton and T.M. Tritt, Review of Scientific Instruments, 72 3129 (2001)
6. A.L. Pope, B. Zawilski and T.M. Tritt, Cryogenics, 41 725 (2001)
7. S. Ouardi, et al., Phys. Rev. B, 82 085108 (2010)
8. J. Yang, et al., Adv. Fun. Mat., 18 2880 (2008)
9. D.G. Cahill, S.K. Watson and R.O. Pohl, Phys. Rev. B, 46 6131 (1992)

Mater. Res. Soc. Symp. Proc. Vol. 1490 © 2013 Materials Research Society
DOI: 10.1557/opl.2013.400

# Erbium boride composites with high ZT values at 800 K

Frederick C. Stober[1,2] and Barbara R. Albert[1]

[1]Eduard-Zintl-Institute of Inorganic and Physical Chemistry, Technische Universität Darmstadt, Petersenstr. 18, 64287 Darmstadt, Germany
[2]Bosch Solar CISTech GmbH, Münstersche Str. 24, 14772 Brandenburg an der Havel, Germany

## ABSTRACT

Single phase erbium borides $ErB_2$, $ErB_4$, and $ErB_{12}$ show Seebeck coefficients and power factors with absolute values that are significantly lower than those of a stable Er-B multi phase composite obtained through high temperature solid-solid reaction from the elements (molar ratio Er:B = 1:6). According to quantitative Rietveld analysis the composite consists of erbium diboride (1 %), tetraboride (83 %), and dodecaboride (16 %), and the measurement of the electrical conductivities, Seebeck coefficients, and thermal conductivities leads to ZT values as high as 0.53 at 830 K. Such refractory materials can be used for energy conversion in a range of high temperatures that are otherwise difficult to address.

## INTRODUCTION

Boron, boron carbide, and many metal borides are semi-conducting or metallic materials that are thermally very stable and that show interesting properties for thermoelectric applications. Elemental boron was reported to show Seebeck coefficients as high as 900 μV/K at room temperature [1], but due to its low electrical conductivity the ZT values are very small. As reported so far, the class of borides and boride silicides has no representative with ZT values higher than 0.25 [2]. Dresselhaus et al. have explained that significant increases in ZT can be expected when nano-structured materials are prepared [3]. But, for high-temperature applications it is often difficult to ensure the stability of nanoscale structures due to the high mobility of atoms at temperatures close to the melting or decomposition temperature. Agglomeration and phase separation can be expected to occur for non-refractory materials in thermoelectric generators. Now, we synthesized several composites of borides that are crystalline and micro- or nano-structured, and tested them for their thermoelectric properties. In addition, the corresponding binary borides were prepared as bulk materials [4]. Since such composites form at temperatures above 2000 K from the elements they can be expected to remain very stable under possible application conditions between 600 and 900 K. Here, we will report on the high-temperature thermoelectric properties of an Er-B composite that consists of three phases, and we will compare it to the single phase binary borides, $ErB_2$, $ErB_4$, and $ErB_{12}$. These are compounds well-known from the Er-B phase diagram, and they are stable up to temperatures as 1912 K ($ErB_2$), 2227 K ($ErB_4$), and 1742 K ($ErB_{12}$) [5]. The only other binary known compound in the Er-B system, $ErB_{66}$, was not observed to form under synthesis conditions described in this work. The existence of erbium hexaboride is controversial, according to literature. Here, $ErB_6$ was not obtained.

**EXPERIMENT**

Samples of the composites were synthesized from elemental erbium (720 mg) and boron (279 mg), according to a molar ratio of erbium to boron of 1:6. Powders were mixed and sintered under argon in tantalum tubes in an induction furnace (Fa. Hüttinger, Freiburg) for five hours at 2100 K. Sintered products were ball-milled for fifteen minutes and pressed into boron nitride crucibles in tantalum tubes to be reheated for another five hours at 2300 K. Samples of the binary erbium borides were synthesized under similar conditions, using the appropriate ratios of Er to B.

X-ray powder diagrams were recorded at room temperature using a STOE StadiP diffractometer (Fa. Stoe, Darmstadt) in transmission geometry with monochromatic Cu radiation ($\lambda$ = 154.056 pm) and flat plate sample holders. The quantitative Rietveld analysis was performed with the program package TOPAS (Fa. Bruker, Karlsruhe).

Thermoelectric properties were investigated with a LSR-3 four-probe instrument (Fa. Linseis, Selb) to measure electrical conductivities and Seebeck coefficients between room temperature and 1050 K, and with a Laserflash instrument LFA 1000 (Fa. Linseis, Selb) to measure thermal diffusivities and conductivities between room temperature and 1050 K.

Homogeneity and the absence of large pores were checked in a scanning electron microscope (SEM) using energy-dispersive spectroscopy (EDS) and performing elemental mapping for boron and erbium.

**RESULTS AND DISCUSSION**

The reaction of erbium and boron (1:6) at temperatures above 2000 K resulted in sintered samples of a composite with a density of 3.8 g/cm$^3$. According to X-ray powder diffractometry (Fig. 1) these samples consist of three crystalline phases, erbium diboride, tetraboride and dodecaboride. No amorphous phase could be detected in considerable amounts (the background of the diffraction data is attributed to the acetate foil used in the sample holder). In independent syntheses single phase binary borides were also prepared as reference materials and sintered under similar conditions. The densities of all sintered samples were approx. 70% of the theoretical densities which is typical for refractory borides.

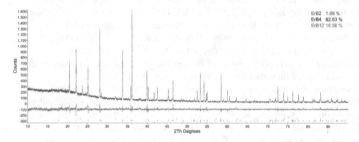

Fig. 1 X-ray powder pattern of a mixture of crystalline ErB$_2$, ErB$_4$, and ErB$_{12}$ (Experimental is data shown in blue, Rietveld fit in red, difference curve in grey)

There ratio of ErB$_2$ to ErB$_4$ to ErB$_{12}$ in the composite (1 %: 83 %: 16 %) was determined with Rietveld analysis using the structure models for the three erbium borides that are known from literature. Experimental and calculated traces fit very well (R$_P$ = 8.27). According to SEM/EDS the distribution of erbium and boron in the sample is homogeneous on a micrometer scale. Measurements of the electrical conductivity for the Er-B composite are shown in figure 2 (left) and compared with the binary borides. It is obvious that the electrical conductivities of the multi phase product are significantly higher than those of the boron-rich reference materials, but lower than those of ErB$_2$ and decreasing with temperature. The Seebeck coefficients of the composite show higher absolute values than those of the single phase binary borides (Fig. 2, right). All of the samples except for ErB$_{12}$ are n-type conductors. A detailed understanding of the temperature dependence of the Seebeck coefficients of both the binary compounds and the composite is yet to be developed.

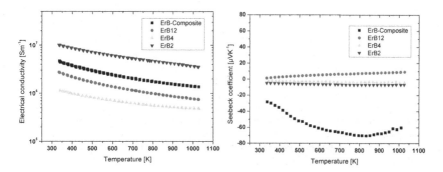

Fig. 2 Electrical conductivities (left) and Seebeck coefficients (right) of the Er-B composite and three binary borides for comparison

When calculating the power factors for the composite a promising value higher than 4 W/(K$^2$m) was found at 830 K. In comparison, the binary borides exhibited values at 830 K as low as 0.018 W/(K$^2$m) for the diboride, 0.006 W/(K$^2$m) for the tetraboride, and 0.05 W/(K$^2$m) for the dodecaboride. Thus, thermal diffusivities and thermal conductivities (Fig. 3, left) were determined for the Er-B composite, to allow for the determination of ZT values for this material (Fig. 3, right). These ZT values range from 0.05 at 350 K up to 0.53 at 830 K, then decreasing again towards 0.4 at 1050 K.

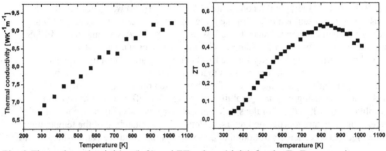

Fig. 3 Thermal conductivities (left) and ZT values (right) for the Er-B composite

## CONCLUSIONS

It was shown that a stable composite of three well-defined, crystalline binary borides can be obtained from reacting erbium and boron at high temperatures. The composite is a n-type conductor and exhibits ZT values that are surprisingly high and that cover an interesting temperature range where not many other thermoelectric materials are known to exist. A corresponding p-type material that is necessary to build a thermoelectric generator for conversion of waste heat into electricity might originate from metal - boron carbide systems that are currently investigated in this context. Further investigations will aim at a better densification of the samples and a more detailed understanding of the temperature dependence of the electrical conductivities and the Seebeck coefficients.

## ACKNOWLEDGMENTS

We would like to thank the Bundesministerium für Bildung und Forschung for financial support under the program ThermoPower – Strom aus Wärme, IN-TEG.

## REFERENCES

1. M. Takeda, T. Fukuda, F. Domingo, and T. Miura, *J. Solid State Chem.* **177**, 471 (2004).
2. T. Mori, T. Nishimura, *J. Solid State Chem.* **179**, 2908 (2006).
3. M. S. Dresselhaus, Y.-M. Lin, S. B. Cronin, O. Rabin, M. R. Black, G. Dresselhaus, in T. M. Tritt (Ed.) Semiconductors and Semimetals, Recent Trends in Thermoelectric MaterialsII, Vol. 71, Academic Press, San Diego, CA, 2001, p. 1-121.
4. F. C. Stober, PhD thesis, Technische Universität Darmstadt, 2012.
5. P. Liao, K. Spear, M. Schlesinger, *J. Phase Equil.* **17**, 326 (1996).

Mater. Res. Soc. Symp. Proc. Vol. 1490 © 2013 Materials Research Society
DOI: 10.1557/opl.2013.173

# Laser Flash Analysis Determination of the Thermal Diffusivity of Si/SiGe Superlattices

Anthony L. Davidson III[1], James P. Thomas[1], Terrance Worchesky[2], Mark E. Twigg[1], and Phillip E. Thompson[2]
[1]Naval Research Laboratory, 4555 Overlook Ave, Washington DC
[2]Retiree Naval Research Laboratory, 4555 Overlook Ave, Washington DC
[3]University of Maryland Baltimore County, 1000 Hilltop Circle, Baltimore MD

## ABSTRACT

Applications that produce a large amount of heat, such as combustion engines, can benefit from high temperature thermoelectrics to reduce the amount of energy lost. Superlattice (SL) structures have shown reduced thermal conductivity at room temperature and below, suggesting applicability at high temperatures may be possible. This reduction could greatly increase the thermoelectric figure of merit. The Si/SiGe material system is studied here for high temperature application. Two growth temperatures of 300 C and 500 C are examined. Two superlattice periods were studied (8 nm and 20 nm) to determine the effects of lattice spacing on thermal conductivity. Laser Flash Analysis is applied to determine the thermal diffusivity, hence thermal conductivity, from 100 C to 500 C. Thermal diffusivity was found to be an order of magnitude lower than the constituent alloy at 100 C. Superlattice spacing and growth temperature showed little effect on the diffusivity within the error of this measurement.

## INTRODUCTION

Converting heat into electrical energy through thermoelectrics has the promise of scavenging waste heat in any combustion process. Automotive combustion engines lose 63% of the energy in the fuel to heat. Equation 1 gives the efficiency ($\eta$) of a thermoelectric device.

$$\eta = \frac{T_1 - T_2}{T_1} \frac{M - 1}{M + \frac{T_2}{T_1}}$$
$$M = (1 + ZT_m)^{1/2}$$

(1)

$T_1$ is the heat source temperature, $T_2$ is the heat sink temperature, $T_m$ is the average of sink and source temperatures, and Z is the figure of merit of the device.[1] The dimensionless figure of merit is normally quoted and is defined as ZT. If the heat source is set at 600 C and the heat sink at 200 C a ZT of 3 gives an efficiency of ~28.57%. This value of ZT is generally considered desirable in thermoelectric generation. SiGe alloy has been used in applications at this temperature range and has a figure of merit of approximately 0.4 giving an efficiency of ~4.75%.

The figure of merit is controlled by the total thermal conductivity ($\kappa_L + \kappa_e$), electrical conductivity ($\sigma$), and the Seebeck coefficient (S) as shown in equation 2.

$$Z = \frac{S^2 \sigma}{\kappa_L + \kappa_e}$$

(2)

Superlattice (SL) structures combine the nanostructuring and alloying techniques to reduce the thermal conductivity of the lattice. Silicon-germanium $(Si/Ge)^2$ and silicon-alloy $(Si/Si_{1-x}Ge_x)^3$ superlattices have shown that this technique results in equal or lower thermal conductivity in the temperature range of -200 C to 127 C compared to the constituent bulk alloy. In this study Si/SiGe SL structures are studied at high temperatures (100 C to 500 C) for application to power harvesting.

**EXPERIMENT**

The superlattices in this study are grown using the Molecular Beam Epitaxy (MBE) technique. Silicon and germanium are grown using electron beam sources. Antimony dopants are grown using Knudsen cells. Deposition rates of silicon and germanium are 1 A/s and .25 A/s respectively giving precise control of the layer thickness in each step. The ratio of layer thickness between Si and SiGe was keep at 1:1 for all structures. Samples were grown on low doped Si substrates 1-10 ohm sq. The alloy ratio of SiGe for all samples was kept at 80% Si to 20% Ge. The alloy is degenerately doped at $5 \times 10^{19}$ cm$^{-3}$. Table I gives a summary of differences between samples examined here. Two parameters are studied here, the substrate temperature during growth and the period of the superlattice.

**Table I.** Samples in this study.

|  | Dopant Type Of SiGe | Substrate Temperature (C) | Superlattice Period (nm) | Number of Layers | Total Thickness (Micron) |
|---|---|---|---|---|---|
| TE 1 | Sb | 500 | 8 | 250 | 2 |
| TE 2 | Sb | 500 | 20 | 100 | 2 |
| TE 3 | Sb | 300 | 8 | 125 | 1 |
| TE 4 | Sb | 300 | 20 | 50 | 1 |

Growth temperature must be chosen carefully to reduce diffusion of impurities in the silicon matrix (Ge, B, P) while maintaining crystallinity of the structure. Conformation of crystallinity is performed by examining structures with a Transmission Electron Microscope (TEM). Additionally, TEM confirms the desired superlattice spacing in the structures. Figure 1 gives images of two structures at the chosen growth temperatures.

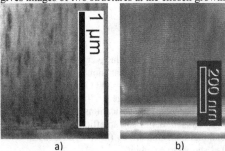

a)                              b)

**Figure 1.** a) TE 4 grown at 300 C. b) TE 1 grown at 500 C

The materials grown at 300 C are shown to be poly crystalline in nature, while the 500 C growth temperature shows high crystallinity. In all structures the lattice spacing is confirmed, however rippling occurs in high temperature samples due to strain from lattice mismatch.

Thermal characterization of superlattice samples is achieved through the use of Laser Flash Analysis (LFA). This technique applies a heat pulse generated by a laser flash on the front side of the sample while measuring the black body radiation from the rear surface vs. time. Thermal diffusivity is then determined by applying the heat flow equation with the boundary conditions show in equation 3 to match the observed temperature rise at the rear surface of the sample.

$$\frac{\partial^2 u(x,t)}{\partial x^2} = \frac{1}{\alpha(x)} \frac{\partial u(x,t)}{\partial t}$$

$$\frac{\partial u}{\partial t}\bigg|_{x=0} = -\frac{q(t)}{\rho C_{\text{tot}}} + v_x u(0,t) \tag{3}$$

$$\frac{\partial u}{\partial t}\bigg|_{x=d} = -v_x u(d,t)$$

In these equations $\alpha(x)$ is the thermal diffusivity, $\rho$ is the density of the material, $C_{lat}$ is the lattice heat capacity, $v_x$ is a radiation heat flow fitting parameter, d is the sample thickness, and u is the temperature difference from ambient. In the single layer model $\alpha(x)$ is a constant, while in the two layer model $\alpha(x)$ has the values of the substrate and thin film with a discontinuity at the growth junction. This equation takes into account the pulse width of the flash ($q(t)$) and radiation heat loss ($v_x u(d,t)$) from the front and rear surface.[4] Absorption and radiation on the respective sides of the sample is facilitated by a graphite coating that has to be corrected for in these measurements. The corrections for the graphite layers are achieved by the modification of the pulse width of the laser and were found to have the same result as using a three layer model by Lim et al. [5] In this study the pulse width is modeled as

$$q(t) = q \frac{t}{\tau^2} e^{-\frac{t}{\tau}} \tag{4}$$

where q is the energy per unit area, and $\tau$ is a pulse width correction that is determined by measuring a silicon sample simultaneously prepared with the SL samples where a single layer model with known diffusivity is applied. This pulse width is then used in a two layer model to determine the SL diffusivity. Thermal conductivity is model is related to the diffusivity of a material through equation 5

$$\kappa = \rho C_{tot} \alpha \tag{5}$$

where $\alpha$ is the diffusivity, $\rho$ is the density and $C_{tot}$ is the total heat capacity of the electrons and phonons. For a superlattice material the average density and heat capacity is used to determine the corresponding thermal conductivity.

**DISCUSSION**

Figure 2 shows the detector voltage versus time of a silicon sample (red circles) and the 8 nm period superlattice (blue squares) on silicon at 500 C. The SL maximum has been normalized to the silicon maximum since the diffusivity only depends on curvature. The SL clearly increases the amount of time for temperature to reach a maximum resulting in a reduction in the apparent diffusivity in a single layer model approach. The curvature modification is not drastic due to the ratio of SL thickness (2 microns) to the substrate thickness (400 microns). Applying a two layer model, the diffusivity of the SL is then extrapolated using the know values of the silicon substrate.

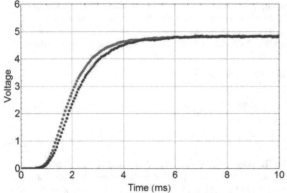

**Figure 2.** Detector Voltage vs. Time of silicon (circles) and 8nm period SL (squares) at 500 C.

The extrapolated SL lattice diffusivities all show values between 1-6 mm²/s. The diffusivity for an alloy of Si.$_9$Ge.$_1$, which is the composition if the germanium were to completely diffuse, is approximately 55 mm²/s at room temperature. Since the diffusivity of the superlattice is an order of magnitude lower that the bulk alloy, diffusion of the Ge in the SiGe layers is not occurring at the high temperatures. This allows the superlattice structures to be applicable over a wide range of temperatures.

Figure 3 shows the diffusivity results of the 8 nm SL (circles) and the 20 nm SL (squares) grown at 500 C. The error bars are one standard deviation of the measured data. The average values show the difference in diffusivity between the structures is minimal. The 20 nm SL shows a slight increase in diffusivity with temperature, while the 8 nm SL shows a constant value within the error of the experiment. Diffusivity of SL samples grown at 300 C show constant values within error limits of the system as well.

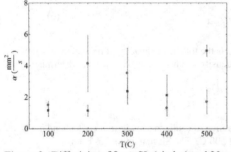

**Figure 3.** Diffusivity of 8 nm SL (circles) and 20 nm SL (square) at 500 C.

Figure 4 compares the 300 C (circles) and 500 C (squares) 8 nm SLs. The substrate temperature also shows little effect on diffusivity between these two samples. Electrical characteristics would have to be performed to determine the best configuration for thermoelectrics since dopant diffusivity is not examined by this technique.

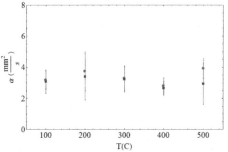

**Figure 4.** Diffusivity of 8 nm SL at a growth temperature of 300 C (circles) and 500 C (squares).

The thermal conductivity is then found using equation 5 with the average heat capacity and density of the layers. In such structures phonon zone folding will also affect heat capacity resulting in the diffusivity conversion as a maximum value. Figure 5 shows the results of the 8 nm SLs compared with Boltzmann Transport Equation under the relaxation time approximation results for bulk $Si_9Ge_1$, and for the case of the SL without zone folding and with zone folding effects.[6] The SL samples exhibit lower thermal conductivity due to the increase in alloy scattering in the $Si_8Ge_2$ layers and the zone folding of phonons under the imposed periodicity. The blue curve shows the expected thermal conductivity if phonon folding does not occur in the SL structures. The black curve is the expected thermal conductivity with phonon folding effects and agrees with measured data. One would expect the polycrystalline material to have a lower thermal conductivity due to increased scattering. Comparison is made to Huxtable et al.[3], who found that a $Si_9Ge_1$ thin film had a thermal conductivity of ~10 W/m K at 50 C, suggesting a stronger scattering mechanism not accounted for in the model. This result would mean merger reductions of thermal properties and require investigation of a SiGe thin film at 1 micron thickness. The authors also found that an increase in SL period for the structures increases thermal conductivity. Reduction of error in measurements is required to determine this effect in our samples.

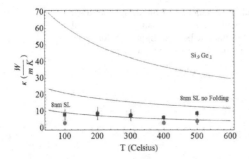

**Figure 5.** Thermal conductivity vs. Temperature. The solid lines are the model predictions. The circles show the data for the 8nm SL at a growth temperature of 500 C and the squares the 8 nm SL at 300 C.

## CONCLUSIONS

In this study we have examined the effects of growth temperature and lattice spacing on thermal conductivity in Si/SiGe SL structures. Results from LFA measurements for thermal diffusivity of superlattice materials show a reduction of an order of magnitude compared with the bulk constituent alloy. Superlattice spacing and growth temperature showed no effect within the error of this measurement. Further LFA studies must be performed to reduce error in the measurement, determine the thermal conductivity of a composite alloy thin film, and to thoroughly determine spacing and substrate effects. Electrical characterization must also be performed to determine the best growth temperature for thermoelectric devices.

## REFERENCES

1.   D. M. Rowe, *CRC handbook of thermoelectrics*. (CRC Press, Boca Raton, FL, 1995).
2.   S. M. Lee, D. G. Cahill and R. Venkatasubramanian, Applied Physics Letters **70** (22), 2957-2959 (1997).
3.   S. T. Huxtable, A. R. Abramson, C. L. Tien, A. Majumdar, C. LaBounty, X. Fan, G. H. Zeng, J. E. Bowers, A. Shakouri and E. T. Croke, Applied Physics Letters **80** (10), 1737-1739 (2002).
4.   J. A. Cape and G. W. Lehman, Journal of Applied Physics **34** (7), 1909-& (1963).
5.   K. H. Lim, S. K. Kim and M. K. Chung, Thermochimica Acta **494** (1-2), 71-79 (2009).
6.   G. P. Srivastava, *The physics of phonons*. (A. Hilger, Bristol ; Philadelphia, 1990).

Mater. Res. Soc. Symp. Proc. Vol. 1490 © 2013 Materials Research Society
DOI: 10.1557/opl.2012.1731

# Enhancement in Thermoelectric Figure-of-merit of n-type Si-Ge Alloy Synthesized Employing High Energy Ball Milling and Spark Plasma Sintering

Sivaiah Bathula[1, 2], M. Jayasimhadri[2], Ajay Dhar[1], M. Saravanan[1], D. K. Misra[1], Nidhi Singh[1], A. K. Srivastava[1], R. C. Budhani[1]

[1]CSIR-Network of Institutes for Solar Energy, CSIR-National Physical Laboratory, Dr. K. S. Krishnan Marg, New Delhi -110012, India
[2]Department of Applied Physics, Delhi Technological University, Delhi, India

ABSTRACT

In the present study, we report the enhancement in figure-of-merit (ZT) of nanostructured *n-type* Silicon-Germanium ($Si_{80}Ge_{20}$) thermoelectric alloy synthesized using high energy ball milling followed by spark plasma sintering (SPS). After 90 h of ball milling of elemental powders of Si, Ge and P (2 at.%), a complete dissolution of Ge in Si matrix has been observed forming the nanostructured n-type $Si_{80}Ge_{20}$ alloy powder. X-ray diffraction analysis (XRD) confirmed the crystallite size of the host matrix (Si) to be ~7 nm and also indicated the formation of an additional phase of SiP nano-precipitates after SPS. HR-TEM analysis revealed that the nano-grained network was retained post-sintering with a crystallite size of size of 9 nm and also confirmed the SiP precipitates formation with a size of 4 to 6 nm. As a result, a very low thermal conductivity of ~2.3W/mK at 900°C has been observed for $Si_{80}Ge_{20}$ alloy primarily due to scattering of phonons by nanostructured grains and nano-scaled SiP precipitates which further contribute to this scattering mechanism. Electrical conductivity values of SiGe sintered alloy are slightly lower to that of reported values in literature. This was attributed to the formation of SiP which creates a compositional difference between the grain boundary region and the grain region, leading to a chemical potential difference at interface and the grain region. Figure-of-merit (ZT) of n-type $Si_{80}Ge_{20}$ nanostructured alloy was found to be ≈1.5 at 900°C, which is the highest reported so far at this temperature.

*Keywords:* Mechanical alloying, Electron Microscopy, Si-Ge thermoelectric alloy, Spark Plasma Sintering, Seebeck coefficient, thermal conductivity, Nanostructured Interfaces

## INTRODUCTION

The direct conversion of heat energy into electricity based on thermoelectric effect without moving parts is attractive alternative for many applications in power generation [1]. The performance of a thermoelectric device depends on its *figure-of-merit (ZT)*, a dimensionless quantity of the material defined as $ZT = (\alpha^2\sigma T/\kappa)$, where $\alpha$, $\sigma$, T and $\kappa$ are Seebeck coefficient, electrical conductivity, temperature and thermal conductivity respectively. The optimization of ZT clearly demands high thermoelectric power and electrical conductivity, but a small thermal conductivity. Improving the performance of a thermoelectric materials or devices mainly involve in controlling the motion of phonons, which carry most of the heat, and electrons, which carry the electric current and some of the heat [2]. Silicon-Germanium (Si-Ge) alloys have been extensively used in deep space missions conducted by NASA. Synthesis of polycrystalline silicon germanium alloys was

---

[1]Corresponding author: adhar@nplindia.org, rcb@nplindia.org
Tel.: +91-11- 4560 9455, Fax.: +91-11-45609310

described in 1981, and the decrease in thermal conductivity related to the phonon scattering and also with smaller grain size was tracked [3]. The current challenge for nanostructured thermoelectric system is how to control the electron scattering at interfaces between randomly oriented grains leading to a concurrent reduction of electrical conductivity.

In the present study, we have employed ball milling technique for nanostructuring and for sintering of Si-Ge alloys we have used a rapid sintering such as spark plasma sintering (SPS). The shorter sintering cycle of SPS was highly beneficial for retaining the nanostructure after sintering. The retained nano-grained network and introduced SiP phase formation after SPS enhanced the power factor and *figure-of-merit (ZT)* significantly.

## EXPERIMENTAL DETAILS

Si (99.99%), Ge (99.99%) and phosphorous (99.9%) powders were ball milled in a *Fritsch* make *Pulverisette-4* ball mill for 90 h. The vial of the ball-mill and the milling media were made up of stainless steel. The weight ratio of ball-to-powder was maintained at 20:1 and milled in argon atmosphere with a milling speed of 400 rpm for 5 min on and off cycles, respectively. In order to prevent the re-welding and promote the fracturing of powder particles, 2 wt.% of stearic acid was added as a process control agent. Prior to consolidation and sintering, ball milled nanostructured alloy powders are kept in glove box *(make: Mbraun)* to minimize oxidation and other atmospheric contamination. Crystallite size measurements are carried out using an X-ray diffractometer *(model: D8 Advanced, make: Bruker)*. In order to calculate the final crystallite size of milled powders, full width half maxima (FWHM) values were subtracted for instrumental broadening and $K\alpha_2$ corrections. Nanostructured alloy powders of Si-Ge subsequently consolidated and sintered under vacuum (~4 Pa) using SPS *(model: SPS 725, make: Dr. Sinter, Japan)* at a pressure of 60 MPa, at 1150°C for 3 min. using graphite die and punches. The heating rate was maintained at 300°C per min. The surface morphology of ball milled powders and sintered pellets were studied using scanning electron microscope *(SEM model: EVO MA10, make: Zeiss)*. The densities of the composites have been measured using conventional Archimedes principle. High resolution transmission electron microscope *(HRTEM, Model: Technai G2 F 30 STWIN, FEG source, electron accelerating voltage 300 kV)* was used for milled powders and SPS derived samples to study the morphology of matrix and the interfaces of alloys. Thermal analyzer *(make: Linsysis model: Laser Flash)* was used to measure the thermal conductivity and Seebeck coefficient, resistivity measurement *(make: Ulvac. Inc., model: ZEM-3)* was used to measure the ZT.

## RESULTS AND DISCUSSION

Fig. 1(a-c) shows the X-ray diffraction (XRD) patterns for (a) un-milled (blend of as-received Si, Ge and P powders), (b) milled for 90 hours and (c) SPS derived n-type SiGe alloy, respectively. It is apparent from Fig. 1(a & b) that after 90 hours of ball-milling, Ge was completely dissolved in Si matrix and the XRD peaks show a significant broadening. Also, it is well known that the impact and shear forces of milling media leads to increase in the internal energy of the system. Beyond a critical value, however, a reduction of energy takes place by forming sub-grains and high angle grain boundaries over the entire volume of milling powder leading to nanostructures [4]. The average crystallite size calculated using Williamson-Hall method [5] was observed to be about 7 nm after 90 hours of ball-milling.

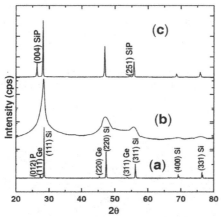

Fig. 1 X-ray powder diffraction pattern of n-type $Si_{80}Ge_{20}$ alloy (a) un-milled (b) milled for 90 hours in argon atmosphere and (c) after spark plasma sintering at 1150°C for 3 minutes under 60 MPa pressure.

High resolution transmission electron microscopy (HRTEM) derived micrographs of 90 hours ball-milled SiGe powder samples exhibited an average particle size distribution between 50 to 80 nm. However, the atomic scale images reveal the individual randomly oriented crystallites with an average crystallite size of about 7nm. The inter-planar spacing of 0.316 nm and 0.192 nm, corresponding to Si (hkl) indices of (111) and (220), respectively, are clearly seen in Fig. 2(b).

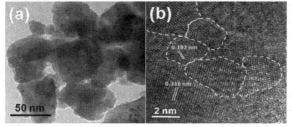

Fig. 2 HRTEM images of $Si_{80}Ge_{20}$ powder ball-milled for 90 hours in argon atmosphere, showing (a) nanocrystalline aggregates in individual particles, (b) nanocrystallites oriented in different directions with different inter-planar spacings of Si, and crystallite size distribution as measured from the microstructure ranging between 2 to 14 nm with average crystallites of size about 7 nm. White dotted encircled regions correspond to the size of individual nano-crystallites with ultra-fine short length boundaries.

A detailed characterization was performed on sintered SiGe alloy employing HR-TEM to reveal the nanoscale features which emerge after SPS. At an atomic scale, it is observed that the nanocrystallites are decorated with large number of grain boundaries and also nano-grained microstructure with randomly oriented nanocrystallites with an average size of ~9 nm was delineated as shown in Fig. 3(a), which are roughly the same as the crystallite size the ball-milled SiGe alloy powders. A large sized nano-crystallite of size ~12nm with oriented planes (hkl: 111) has been marked with an arrow and depicted in Fig. 3(b). In the present study, with the addition of 2 at.% P to the SiGe alloy resulted in the formation of SiP as shown in Fig. 3(c) and this was evidenced by XRD analysis as well (Fig. 1(c)). The formation of SiP phase after hot-pressing in P-doped SiGe alloys has also been reported earlier [6] also on similar composition alloys. The SiP phase having an inter-planar spacing of about 0.169 nm are seen at the grain boundaries and the size of these precipitates size is about 4 – 6 nm.

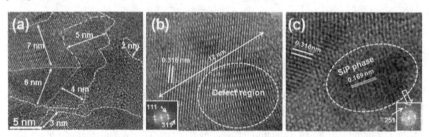

Fig. 3 HRTEM images of spark plasma sintered n-type nanostructured alloys, (a) distribution of different sized nanocrystallites with boundaries marked with white dotted lines, (b) a large sized nanocrystal of about 12 nm with 0.316 nm inter-planar spacing and defect layers at lattice scale , and (c) SiP phase with inter-planar spacing of 0.169nm in the matrix of $Si_{80}Ge_{20}$. Insets in (b) FFT of $Si_{80}Ge_{20}$ matrix showing the planes (111) and (311), and (c) FFT of SiP phase showing the plane (251)

The temperature dependence of thermoelectric properties of the sintered n-type SiGe nanostructured alloys are shown in Fig.4. It well known that the thermal conductivity is greatly dependent on the crystallite size [7]. The thermal conductivity was observed to be ~2.3W/mK at 900°C (Fig. 4(a)), which is so far the lowest value reported for SiGe alloys at this temperature. This low value of thermal conductivity can be attributed to the scattering of phonons of larger spectrum of wavelengths (2-300nm for SiGe alloys) by high density grain boundaries [8] and the retained nano-grained network has an average crystallite size of about ~9nm. In the present study, apart from the dense grain boundaries, nano-precipitates of SiP having a size of about 4 – 6 nm (Fig. 3 (c)) could also act as additional scattering centers for long and mid-wavelength phonons in the SiGe matrix.

The electrical conductivity of n-type SiGe alloys is displayed in Fig. 4(b). The electrical conductivity shows a rather sharp drop with increasing the temperature from room temperature and reaches a value of ~3.6 x $10^4$ S/m at 900°C. This value, however, is lower than the reported value by Wang et al. [9] on hot-pressed SiGe alloys. The lower magnitude of electrical conductivity due to the formation of SiP nano-precipitates (4 - 6nm) along the grain boundaries (XRD in Fig. 1(c) and HR-TEM image Fig. 3(c)) which can lead to a compositional difference between the grain boundary region and the grain region.

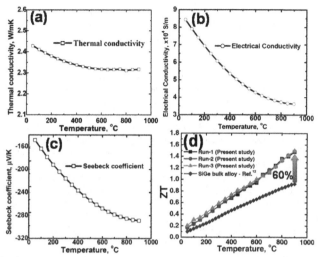

Fig. 4 Temperature dependent thermoelectric properties of n-type SiGe alloy (a) thermal conductivity, (b) electrical conductivity, (c) Seebeck coefficient, and (d) Figure-of-merit (ZT).

This chemical potential difference resulting from such segregation could modulate the phonon scattering mechanism at interface and grain region, thus resulting in lower electrical conductivity [10]. Thus, the reduction in the electrical conductivity, observed in the present case, could additionally be due to electron-interface and electron-defect scattering. Fig. 4(c) which shows the temperature dependence of Seebeck coefficient indicated that increases linearly from temperature to a value of about -290 $\mu$V/K at around $750^\circ$C and then tends to saturate, contrary to the behavior reported earlier by Wang et al. [9] on same composition of SiGe alloys, which shows a peak at around $650^\circ$C and tends decrease to a lower magnitudes at 900°C. The magnitude of Seebeck coefficient, obtained in the present studies is the highest reported value so far in n-type SiGe alloys. Earlier reports on similar composition used hot pressing as the sintering resulted in an average grain size of about 22 nm with a ZT of about 1.3 at 900°C. However, in the present work, ZT was observed to be 1.5 at 900°C (Fig. 4 (d)) with an average grain size of about ~9nm after SPS, leading to a very high interface density. The measurements of thermoelectric properties were repeated several times using the same thermal treatment (reset) of the samples [11], prior to each measurement and the resulting ZT plots, depicted in Fig. 4(d), show that the repeatability was reasonably good. This value of ZT is more than 60% higher than its Radioisotope thermoelectric generator (RTG) [12] and more than 15% higher than the reported record value of n-type $Si_{80}Ge_{20}$ nanostructured alloy [9,13]. Thus, it is concluded that the enhancement in ZT of the present n-type SiGe alloys was primarily due to a lower value of thermal conductivity due to increased phonon scattering by high density of grain boundaries, nano-precipitates of SiP phase along the grain boundaries and defects introduced by ball-milling and rapid sintering by SPS.

## CONCLUSIONS

In summary, the efforts in the present work were successful in achieving a ZT of about 1.5 at 900°C, which is the highest reported value so far on n-type $Si_{80}Ge_{20}$ nanostructured alloys. This high value of ZT was possible due to the nanostructuring leading to nano-scale features, introduced by ball milling, which were retained post-sintering primarily because the nanostructured powders were consolidated with rapid-sintering rates employing SPS. The microstructure indicated nanocrystalline features along with high density of grain boundaries, presence of nano-sized SiP precipitates along the grain boundaries and point defects introduced due to ball-milling. The nanostructuring favorably altered the thermal conductivity and Seebeck coefficient in such a way that the combined effect leads to the enhancement in ZT, which improved to the extent of more than 15%, as compared to the reported record value [9,13] on SiGe alloy of similar composition.

## ACKNOWLEDGEMENTS

This work was financially supported by CSIR-TAPSUN (CSIR-NWP-54) program entitled *"Novel approaches for solar energy conversion under technologies and products for solar energy utilization through networking"*. The authors are grateful to R. C. Anandani, Radhey Shyam and N. K. Upadhyay for their technical and experimental support.

## REFERENCES

1. G. J. Snyder and E. S. Toberer, *Nature Mater.* **7**, 105 (2008).
2. B. Poudel, Q. Hao, Y. Ma, Y. C. Lan, A. Minnich, B. Yu, X. Yan, D. Z. Wang, A. Muto, D. Vashaee, X. Y. Chen, J. M. Liu, M. S. Dresselhaus, G. Chen, and Z. F. Ren, *Science* **320**, 634 (2008).
3. M. Rowe, V. S. Shukla, and N. Savvides, *Nature* **290**, 765 (1981)
4. F. L. Zhang, C. Y. Wang, M. Zhu: *Scripta Materialia*, 2003, 49, 1123.
5. B. D. Cullity, S. R. Stock: *Elements of X-Ray Diffraction* .3rd Ed., Prentice-Hall Inc., 2001, 167.
6. G. H Zhu, H Lee, Y. C Lan, X. W Wang, G Joshi, D. Z Wang, J Yang, D Vashaee, H Guilbert, A Pillitteri, M. S Dresselhaus, G Chen, Z. F Ren, *Phys. Rev. Lett.* **102**, 196803 (2009)
7. Liang Yin, Eun Kyung Lee, Jong Woon Lee, Dongmok Whang, Byoung Lyong Choi, *Appl. Phys. Lett.* **101**, 043114 (2012)
8. Nikhil Satyala and Daryoosh Vashaee, *Appl. Phys. Lett.* **100**, 073107 (2012)
9. X. W. Wang, H. Lee, Y. C. Lan, G. H. Zhu, G. Joshi, D. Z. Wang, J. Yang, A. J. Muto, M. Y. Tang, J. Klatsky, S. Song,1 M. S. Dresselhaus, G. Chen, and Z. F. Ren, *Applied Physics Letters* **93**, 193121 (2008).
10. G. H Zhu, H Lee, Y. C Lan, X. W Wang, G Joshi, D. Z Wang, J Yang, D Vashaee, H Guilbert, A Pillitteri, M. S Dresselhaus, G Chen, Z. F Ren, *Phys. Rev. Lett.* **102**, 196803 (2009)
11. B. A Cook, J. L Harringa, S. H Han and C. B Vinning, *J. Appl. Phys.***78**, 5474 (1995)
12. C. Wood, *Rep. Prog. Phys.* **51**, 459 (1988)
13. B.Yu, M. Zebarjadi, H. Wang, K. Lukas, H. Wang, D. Wang, C. Opeil, M. Dresselhaus, G. Chen, and Z. Ren, *Nano Lett.* **12**, 2077 (2012)

Mater. Res. Soc. Symp. Proc. Vol. 1490 © 2013 Materials Research Society
DOI: 10.1557/opl.2013.52

# Thermoelectrical properties of α phase and γ phase $Na_xCo_2O_4$ ceramics prepared by spark plasma sintering method

Natsuko Mikami[1], Keishi Nishio[1], Koya Arai[1], Tatsuya Sakamoto[1], Masahiro Minowa[2], Tomoyuki Nakamura[2], Naomi Hirayama[1], Yasuo Kogo[1] and Tsutomu Iida[1]
[1]Department of Materials Science and Technology, Tokyo University of Science, 2641 Yamazaki, Noda-shi, Chiba 287-8510
[2]SWCC Showa Cable Systems Co., Ltd., LTD, 4-1-1 Minami-Hashimoto Chuo-Ku Sagamihara, Kanagawa, Japan

## ABSTRACT

The thermoelectrical properties of α and γ phases of $Na_xCo_2O_4$ having different amounts of Na were evaluated. The γ $Na_xCo_2O_4$ samples were synthesized by thermal decomposition in a metal-citric acid compound, and the α $Na_xCo_2O_4$ samples were synthesized by self-flux processing. Dense bulk ceramics were fabricated using spark plasma sintering (SPS), and the sintered samples were of high density and highly oriented. The thermoelectrical properties showed that γ $Na_xCo_2O_4$ had higher electrical conductivity and lower thermal conductivity compared with α $Na_xCo_2O_4$ and that α $Na_xCo_2O_4$ had a larger Seebeck coefficient. These results show that γ $Na_xCo_2O_4$ has a larger power factor and dimensionless figure of merit, $ZT$, than α $Na_xCo_2O_4$.

## INTRODUCTION

It was reported by Terasaki [1] that the thermoelectric oxide material $Na_xCo_2O_4$ had a high thermoelectric power at high temperatures. $Na_xCo_2O_4$ has a large Seebeck effect and high electrical conduction because of its strongly correlated electronic structure. In addition, it is an intercalation compound and has a layered crystal structure with sodium ions in the Co oxide between layers [3,4]. The inter-lamellar sodium ions suffer losses at random, and this nonstoichiometry causes the properties of $Na_xCo_2O_4$ to change. Additionally, the crystal structure of $Na_xCo_2O_4$ changes in a phase transition when the sodium defects are released from it, and its thermoelectric properties depend on the phase [3-5]. $Na_xCo_2O_4$ has three crystal structures depending on the $x$ value in the system; β-phase (P3 type, $1.1<x<1.2$), γ-phase (P2 type, $1.0<x<1.4$), and α-phase (O3 type, $1.8<x<2.0$) [2]. The P2 type γ-phase $Na_xCo_2O_4$ has a large Seebeck coefficient despite its metallic conductivity. Various studies have reported the thermoelectrical properties of $Na_xCo_2O_4$, however they have been inclined to deal with only the P2-type γ phase. Few have attempted a comparison of the properties with the O3-type α phase or P3-type β phase. Therefore, in this study, we focused on the γ phase and α phase created by the conversion of the sodium content of $Na_xCo_2O_4$ and evaluated the composition and thermoelectric characteristics of powder and sintered samples.

## EXPERIMENT

The $\gamma$-Na$_x$Co$_2$O$_4$ polycrystalline powders were synthesized by thermal decomposition in a metal-citric acid compound [7]. This process yields high-purity, good crystalline materials due to its combination at the molecular level with the solution method. First, citric acid monohydrate (C$_6$H$_8$O$_7$ H$_2$O, purity 99.0%) was completely dissolved into 2-methoxyethanol, after which sodium acetate (CH$_3$COONa, purity 99.0%) and cobalt nitrate hexahydrate (Co(NO$_3$)$_2$ 6H$_2$O, purity 98.5%) was added, weighted with a molar fraction of Na: Co=2.2: 2.0 (batch composition). Next, the solution was stir heated in an oil bath at 333 K for one hour in air and heat-treated in an electric furnace at 723 K for 3 hours. Finally, the provided powder was heat treated at 1173 K for one hour in air after grinding with an alumina mortar. A powdery sample of $\gamma$-Na$_x$Co$_2$O$_4$ was obtained after grinding again. The $\alpha$-Na$_x$Co$_2$O$_4$ polycrystalline powders were fabricated using a self-flux method because it was difficult to synthesize $\alpha$-Na$_x$Co$_2$O$_4$ powder by thermal decomposition. The reason for this difficulty was that the sodium ions volatilized at high temperature, as confirmed by TG-DTA [6], and a high enough Na concentration could not obtained. In the self-flux method, cobalt oxide and sodium hydroxide as a flux were put into an alumina crucible and sealed with an alumina cover in a dry nitrogen atmosphere. Next, the sample was heat-treated at 973 K for 5 hours and cooled to 673 K for 15 hours. It was then crushed. A powdery sample of $\alpha$-Na$_x$Co$_2$O$_4$ was obtained after grinding with an alumina mortar.

The powders were placed in a graphite die and spark plasma sintered (SPS) at 1073 K for 1 minute under 60.0 MPa of uniaxial pressure in a vacuum by a DR. SINTER LAB 515S (Fuji Electronic Industrial Co., Ltd, Japan). The SPS technique enables a dense bulk to be fabricated in a short time. The sintered samples were cut using an ISOMET 4000 (Buehler, USA). The powder samples and bulk were observed by scanning electron microscopy (SEM) (JCM-5100; JOEL LTD., Japan) and characterized by X-ray diffraction with Fe-K$\alpha$ radiation (ULTIMA-4; Rigaku Corp., Japan). The thermoelectric properties of these bulk ceramics were evaluated in terms of their electric conductivity, $\sigma$, Seebeck coefficient, $S$ (ZEM-2; ULVAC-RIKO Inc., Japan), and thermal conductivity, $\kappa$, was analyzed by laser flash (LF) (TC-7000H; ULVAC-RIKO Inc., Japan). The power factor and dimensionless figure of merit $ZT$ was calculated as $PF=S^2\sigma$, $ZT=S^2\sigma/\kappa T$. The open-circuit voltage and thermoelectric power outputs of elements with temperature differences, $\Delta T$, ranging from 373 to 873 K were measured in air (UMTE-1000M; Union Material).

## RESULTS AND DISCUSSION

SPS produced the $\alpha$ phase and $\gamma$ phase bulks at a lower temperature and after a shorter time compared with other sintering techniques, for example, the hot press process. The obtained ceramic had a high bulk density, more than 97.0% of the theoretical X-ray density. The powder and bulk samples were identified as monophasic $\gamma$-Na$_x$Co$_2$O$_4$ and $\alpha$-Na$_x$Co$_2$O$_4$ by XRD. In spite of the molar fraction of Na: Co being 2.2: 2.0 (batch composition), single-phase $\gamma$-Na$_x$Co$_2$O$_4$ ($1.0<x<1.4$) could be obtained because the vapor pressure of Na was very high in the high temperature range used for synthesizing the Na$_x$Co$_2$O$_4$ powder. The heat treatment removed a large amount of Na from the Na$_x$Co$_2$O$_4$ powder. As a result, we obtained pure $\gamma$-Na$_x$Co$_2$O$_4$ powder. Thermal decomposition yielded single-phase $\gamma$-Na$_x$Co$_2$O$_4$. Moreover, the self-flux yielded single-phase $\alpha$-Na$_x$Co$_2$O$_4$. The $\gamma$-Na$_x$Co$_2$O$_4$ and $\alpha$-Na$_x$Co$_2$O$_4$ grains were plate-like in shape. We compared the diffraction patterns of the powder and ceramics with that of a reference and found that the peaks of the (*00l*) phases were more intense than those of the other phases. The *c*-axis of the prepared ceramic was perpendicularly aligned with the uniaxial pressure direction [7]. The XRD pattern of the powder was similar to that of the ceramic because the powder was loaded into the measuring cell under uniaxial pressure. These results indicate that plate-like grains aligned with the *c*-axis direction through the application of uniaxial pressure. Moreover, they confirm that the samples were oriented ceramics. The Na$_x$Co$_2$O$_4$ consisted of Na cations and CoO$_2$ blocks alternately stacked along the *c*-axis, forming a layered structure.

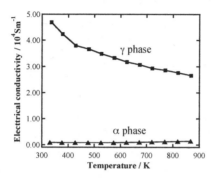

**Figure 1.** Temperature dependence of electrical conductivity of $\gamma$-Na$_x$Co$_2$O$_4$ and $\alpha$-Na$_x$Co$_2$O$_4$.

Figure 1 shows the temperature dependence of the electrical conductivity of the sintered $\gamma$– and $\alpha$-Na$_x$Co$_2$O$_4$ ceramic samples in the direction perpendicular to the applied pressure. The $\gamma$ phase was higher in electrical conductivity than the $\alpha$ phase in the operating temperature range (300 K to 873 K). For this reason, the carrier concentration contributed to electrical conduction; the $\gamma$-Na$_x$Co$_2$O$_4$ (1.0<$x$<1.4) with a high carrier concentration due to many Na defects was a good electrical conductor. With increasing temperature, however, the electrical conductivity of the $\gamma$ phase decreased and behaved like a metallic electrical conductor, because the lattice vibrations became more active due to the loss of Na; this reduced the mobility of the carriers. Generally, the electrical conductivity of ceramics with few carriers relative to metal is influenced by the phonons. On the other hand, the electrical conductivity of the $\alpha$ phase increased with temperature and had a semiconducting behavior. The $\alpha$ phase's tendency to increase in electrical conductivity comes from carriers being added by thermal excitation and the presence of Na defects at high temperatures above 500 K. Moreover, the slight upward trend lasts until 500 K and is affected by phonons acting to limit the carrier mobility. In addition, in the polycrystalline samples, external factors such as the lower carrier mobility due to the crystal grain boundary influence the electrical resistance.

Figure 2 shows the Seebeck coefficient, $S$, as a function of temperature. The Seebeck coefficient was positive for all of the $Na_xCo_2O_4$ samples, and it indicated p-type conductivity. The $\alpha$ phase's Seebeck coefficient increased with temperature, and it was larger than that of the $\gamma$ phase because carrier concentrations due to the small amount of Na defects were lower than in the $\gamma$ phase. Unlike the electrical conductivity, the external dispersion by grain boundary diffusion and defects does not affect the Seebeck coefficient; the coefficient is thus a suitable property for showing the change in carrier levels. In addition, the tendency for the to the Seebeck coefficient of the $\alpha$ phase to increase with temperature inversely relates to the behavior of the electrical resistance due to thermal excitation of the carriers.

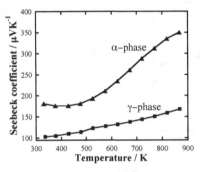

**Figure 2.** Seebeck coefficient of $\gamma$-$Na_xCo_2O_4$ and $\alpha$-$Na_xCo_2O_4$ as a function of temperature.

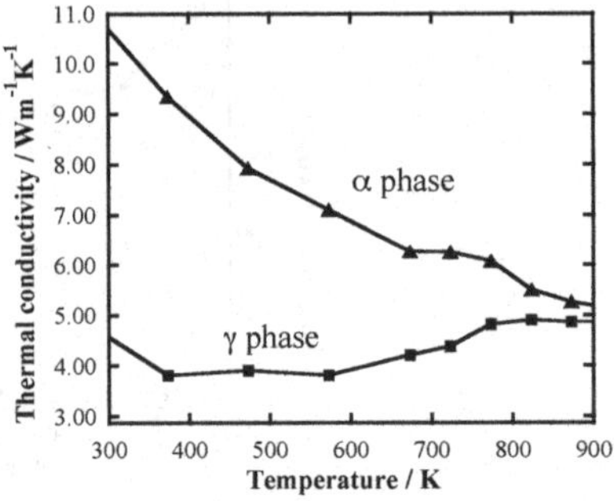

**Figure 3.** Thermal conductivity of $\gamma$-$Na_xCo_2O_4$ and $\alpha$-$Na_xCo_2O_4$ as a function of temperature.

Figure 3 plots thermal conductivity, $\kappa$, as a function of temperature. The $\gamma$ phase had lower thermal conductivity than the $\alpha$ phase, and the $\alpha$ phase decreased with increasing temperature. It is expected that phonon scattering in $\gamma$ phase sample is larger than in $\alpha$ phase sample because the Na defect concentration of $\alpha$ phase is lower than that of $\gamma$ phase.

Figure 4 shows the power factor and dimensionless figure of merit, $ZT$, for two different phases. The power factor and $ZT$ increased with temperature. The power factor and $ZT$ of the $\gamma$-$Na_xCo_2O_4$ sample with high electrical conductivity and low thermal conductivity were higher than those of the $\alpha$-$Na_xCo_2O_4$ sample with the high Seebeck coefficient. The highest $ZT$ was obtained for the $\gamma$ phase ($ZT$=0.125 at 873 K). The carrier thermal conductivity, $\kappa_{el}$, was estimated from the Wiedemann-Franz law, and the thermal lattice thermal conductivity, $\kappa_{ph}$, was obtained from the total thermal conductivity. Figure 5 plots $\kappa_{el}$ and $\kappa_{ph}$ as a function of temperature for $\gamma$ and $\alpha$-$Na_xCo_2O_4$. These results indicate that the thermal conductivity of $Na_xCo_2O_4$ is due in large part to the lattice thermal conductivity, $\kappa_{ph}$. The lattice thermal conductivity, $\kappa_{ph}$, of the $\gamma$ phase was less than that of the $\alpha$ phase (Fig. 5). The reason is that vacancies due to Na loss significantly distort the crystal lattice of $\gamma$-$Na_xCo_2O_4$ and reduce the mean free path of phonons. As a result, the thermal conductivity of $\gamma$-$Na_xCo_2O_4$ was low. On the other hand, the carrier thermal conductivity, $\kappa_{el}$, of the γ phase was higher than in the α phase. Accordingly, the γ phase in which a lot of Na was lost had higher carrier levels than the α phase did.

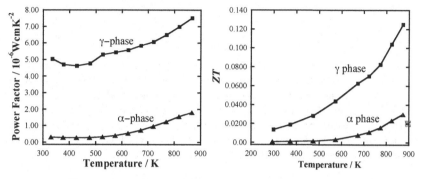

**Figure 4**. Power factor and dimensionless figure of merit ($ZT$) as a function of temperature for γ-$Na_xCo_2O_4$ and α-$Na_xCo_2O_4$.

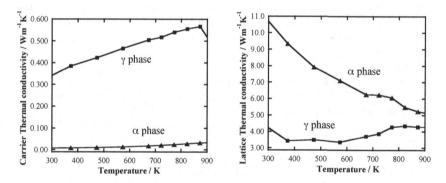

**Figure 5**. Carrier thermal conductivity and Lattice thermal conductivity of γ-$Na_xCo_2O_4$ and α-$Na_xCo_2O_4$ as a function of temperature.

**Table 1.** Maximum values of open-circuit voltage and thermoelectric power outputs of electrode elements ($2*2*8$ mm$^3$) in $\Delta T=500$ K.

| Crystal phase | Open voltage (mV) | Max current (mA) | Max current density (mA/mm$^2$) | Output power (mW) |
|---|---|---|---|---|
| $\gamma$ | 49.9 | 457 | 114 | 5.70 |
| $\alpha$ | 73.6 | 73.1 | 18.3 | 1.39 |

Table 1 shows the measurements of the electrode elements including the open voltage, maximum electric current, highest current density, and maximum electrical power over $\Delta T=500$ K. $\gamma$-Na$_x$Co$_2$O$_4$ had high maximum electrical power. The element made of the $\gamma$ phase had an electric current (max.456.6 mA) more than three times that of the $\alpha$ phase, because of its high carrier level. On the other hand, the element made of the $\alpha$ phase had a high open voltage of 73.3 mV, 1.5 times that of the $\gamma$ phase, because it had a large Seebeck coefficient, on which the opening voltage depends.

## CONCLUSION

$\gamma$-Na$_x$Co$_2$O$_4$ samples were synthesized using thermal decomposition with a metal-citric acid compound, and $\alpha$-Na$_x$Co$_2$O$_4$ samples were synthesized by self-flux processing. The $\gamma$-Na$_x$Co$_2$O$_4$ samples had a high electric current and the $\alpha$-Na$_x$Co$_2$O$_4$ had a large voltage. The different properties were due to the different carrier levels and phonon actions of the samples. Thus, we made it possible to make Na$_x$Co$_2$O$_4$ materials having different electrical characteristics through the adjustment of the Na content. We expect that this advance will have thermoelectric modularization applications.

## REFERENCES

1. I. Terasaki, Y. Sasago and K. Uchinokura, *Phys. Rev. B*, 1997, 56, 12685
2. H. Yakabe, K. Fujita, K. Nakamura and K. Kikuchi, *Proc. 17th Int. Conf. on Thermoelectric*, (Nagoya, Japan, 1998) pp. 567-569.
3. I. Terasaki, Y. Sasago and K. Uchinokura, *Proc.17th Int. Conf. on Thermoelectric*, (Nagaya, Japan, 1998) pp. 564-569
4. H. Yakabe, K. Kikuchi, I. Terasaki, Y. Sasago and K. Uchinokura, *Proc. 16$^{th}$ Int. Conf. on Thermoelectric*. (Dresden, Germany, 1997) pp. 523-527.hi
5. I. Terasaki, Y. Ishii, D. Tanaka, K. Takahata and Y. Iguchi: *Jpn. J. Appl. Phy.* 40 (2001) L65-L67
6. T. Motohashi, E. Naujails, R. Ueda, K. Isawa, M. Karppinen and H. Ya-mauchi, *Appl. Phys. Lett.* 79, 1480 (2001).
7. Keishi Nishio, Kazuma Takahashi, Yusuke Inaba, Mariko Sakamoto, Tsutomu Iida, Kazuyasu Tokiwa, Yasuo Kogo, Atsuo Yasumori, Tsuneo Watanabe , *23rd International Conference on Thermoelectrics-ITC2004 in Adelaide (Australia) Proceedings CD* (Paper No. #97), Released (2005)

Mater. Res. Soc. Symp. Proc. Vol. 1490 © 2013 Materials Research Society
DOI: 10.1557/opl.2012.1732

# Fabrication of Mg$_2$Si bulk by spark plasma sintering method with Mg$_2$Si nano-powder

Koya Arai, Keishi Nishio, Norifumi Miyamoto, Kota Sunohara, Tatsuya Sakamoto, Hiroshi Hyodo, Naomi Hirayama, Yasuo Kogo and Tsutomu Iida

Department of Materials Science and Technology, Tokyo University of Science, 2641 Yamazaki, Noda-shi, Chiba 287-8510, Japan

## ABSTRACT

Mg$_2$Si bulk was fabricated by spark plasma sintering (SPS) nano-powder, and the thermoelectric characteristics of the bulk sample were evaluated at temperatures up to 873 K. A pre-synthesized all-molten commercial polycrystalline Mg$_2$Si source (un-doped n-type semiconductor) was pulverized into powder of 75 μm or less. To obtain nano-sized fine powder, the powder was milled using planetary ball mill equipment under an inert atmosphere. Fine Mg$_2$Si nano-powder with a mean grain size of about 500 nm was obtained. XRD analysis confirmed that no MgO existed in the nano-powder. The fine powder was put in a graphite die to obtain a sintering body of Mg$_2$Si and treated by SPS under vacuum conditions. The resulting Mg$_2$Si bulk had high density and did not crack. However, the XRD analysis revealed a small amount of MgO in it. The thermoelectric properties (electrical conductivity, Seebeck coefficient, and thermal conductivity) were measured from room temperature to 873 K. The microstructure of the sintered body was observed by scanning electron microscopy. The maximum dimensionless figure of merit of a sample made from Mg$_2$Si nano-powder was $ZT = 0.67$ at 873 K.

## INTRODUCTION

Thermoelectric (TE) power generation has recently attracted much interest as a technology for the conversion of waste heat into electric power. The performance of TE materials is usually denoted as dimensionless figure of merit, $ZT = S^2 \sigma T / \kappa$, where $S$ is the Seebeck coefficient, $\sigma$ is electrical conductivity, $T$ is temperature, and $\kappa$ is thermal conductivity. n-type magnesium silicide (Mg$_2$Si) has been studied by many researchers because of its non-toxicity, environmental friendliness, lightness, and comparatively abundance compared with other TE systems. Moreover, Mg$_2$Si has high electrical conductivity, low thermal conductivity, and a high Seebeck coefficient. Mg$_2$Si is a candidate TE material operating in the temperature range from 600 K to 900 K, a range that corresponds to the operating temperatures of industrial furnaces, automobile exhausts, and incinerators [1-5]. Our group has reported that Mg$_2$Si bulk material, consisting of micron-order grains and prepared by spark plasma sintering, had high thermoelectric performance ($ZT$=1.08) [6]. A large number of studies have been reported about synthesis method, TE performance of Mg$_2$Si [7-12]. However, it has been reported that nano-structuring can give rise to large reductions in lattice thermal conductivity due to an increase in phonon scattering at interfaces between grains. As a result, this may raise the $ZT$ of thermoelectric materials [13,14]. In this study, Mg$_2$Si nano-powder was fabricated by a planetary ball milling and Mg$_2$Si bulk was fabricated from the nano-powder by a spark plasma sintering (SPS). The thermoelectric performances of the bulk material were evaluated. The SPS method is useful for preparing polycrystalline material with high density and reduced grain growth. The process is easy and cost effective, and it requires no previous sintering experience. Unlike the conventional powder sintering methods, SPS makes it possible to prepare ceramics at low temperatures and over a shorter duration by charging the intervals between the powder particles with electrical energy and applying a momentary highly energized spark electrical discharge. Moreover, it makes it possible to synthesize metal compounds, ceramics, polymers, and thermoelectric semiconductors [15,16].

## EXPERIMENT

Pre-synthesized polycrystalline $Mg_2Si$ was provided by Yasunaga Corporation, Japan. The pre-synthesized $Mg_2Si$ was prepared from a mixture of high-purity Mg (99.95%) and Si (99.99999%) in the stoichiometric ratio Mg : Si = 2 : 1, and it was heat-treated in an electrical furnace. The polycrystalline bulk $Mg_2Si$ was ground into a powder and sieved to a powder-particle size of 75 μm or less in a dry box filled with Ar gas. Dry ball milling was carried out using planetary ball mill equipment (FRITSCH, P-6) under an Ar atmosphere. $Mg_2Si$ powder was milled in a zirconia pot with a zirconia ball (ball size of 0.5 mm : 0.8 mm : 2 mm = 1 : 1 : 1 in volume fraction) at 300 rpm for 5 h. Spark plasma sintering (SPS) was carried out using a DR. SINTER LAB 515S system (Fuji Electronic Industrial Co., Ltd, Japan). For the SPS, the milled fine powder was placed in a graphite die and heat-treated up to a pre-defined temperature (1113 K) at a rate of 100 K/min, after which it was maintained at that temperature for 10 min under a uni-axial pressure of 30 MPa under vacuum conditions. After attaining and holding the sintering temperature in the SPS process, the electrical power was turned off at the holding temperature, and the sample was cooled to room temperature in the SPS chamber. The sintered body was cut and polished. The relative density of the sample was measured using the Archimedes' method. The XRD patterns of the $Mg_2Si$ nano-powder and bulk were measured using an Ultima IV system (Rigaku Co., Japan). The X-ray fluorescence (XRF) analysis of the $Mg_2Si$ bulk was carried out using a ZSX (Rigaku Co., Japan). A microstructure of the sample was investigated using a scanning electron microscope JCM-5100 system (JEOL Ltd., Japan). The Seebeck coefficient ($S$) and electrical conductivity ($\sigma$) were measured using a standard four-probe method (ULVAC-RIKO, ZEM-2) in a He atmosphere in a temperature range from 300 to 1023K. The thermal conductivity ($\kappa$) was measured using a laser-flash system (ULVAC-RIKO, TC-7000H).

## DISCUSSION

### Analyses of the $Mg_2Si$ powder and bulk

Figure 1 shows the XRD patterns of (a) a JCPDS reference of $Mg_2Si$ [17], (b) the $Mg_2Si$ micro-powder (~75 μm), (c) nano-powder of after ball milling, (d) bulk made with micro-powder, and (e) bulk made with nano-powder. The intense diffraction peaks could be assigned to those of $Mg_2Si$ in all XRD patterns. However, an intense diffraction peak of the Si phase at 28.4° was observed in the XRD pattern of the $Mg_2Si$ nano-powder (b), and Si and MgO (peak positions at 42.9°) peaks were observed in pattern of the bulk made with nano-powder (d). It was thought that $Mg_2Si$ was oxidatively decomposed into Si and MgO or amorphous phases of $SiO_2$ and MgO during the ball milling treatment; oxygen atoms may have moved from the zirconia (ball and pot) into the $Mg_2Si$. In the XRF measurement, 0.003at% zirconium and 5.239at% oxygen

5.239at% were detected in the cross section of a Mg$_2$Si bulk sample made with nano-powder. On the other hand, 3.406at% oxygen was detected in the bulk made from micro-powder.

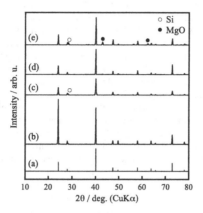

Figure 1. XRD patterns of a JCPDS reference of Mg$_2$Si [17] (a), Mg$_2$Si micro-powder (~75 μm) (b), nano-powder of after ball milling (c), bulk made from micro-powder (d), and bulk made from nano-powder (e).

Figure 2 shows SEM images of Mg$_2$Si powder treated by planetary ball milling (a) and a cross section of a bulk sample fabricated from fine powder by SPS (b). The grain size of the powder (fig. 2 (a)) was less than 3 μm. In regard to the measured particle size distribution, the most common particle size was 400 to 600 nm, and the average particle size was about 500 nm. There were many small particles less than 100 nm in size. The bulk sample (fig. 2 (b)) appeared to be free of any pores or cracks. The relative density of a sample made with nano-powder (99.8%) was higher than that of the sample made with micro-powder (95.4%). It seems that the sintered density depended on the particle size of the raw powder.

(a)                                   (b)
Figure 2. SEM images of Mg$_2$Si powder after ball milling:
(a) Mg$_2$Si powder treated by planetary ball milling and (b) cross
section of the bulk sample fabricated from fine powder by SPS.

65

**Thermoelectric properties of the Mg₂Si bulk**

Figure 3 show the temperature dependence of the Seebeck coefficient ($S$), electrical conductivity ($\sigma$), and thermal conductivity ($\kappa$) of the Mg₂Si bulk sample made from micropowder (~75 μm) [17] and a sample made from nano-powder. In figure 3 (a), the electrical conductivity of both samples decreased and the Seebeck coefficient increased with increasing temperature from 323 K to 673 K. Conversely, in temperature range from 673 K to 873 K, the electrical conductivity increased and the Seebeck coefficient decreased with increasing temperature. The thermoelectric properties of the two samples were similar. The Seebeck coefficients of these materials increased with decreasing electrical conductivity because the electrical conductive mechanism was metallic. On the other hand, in the high temperature range between 323 and 723 K, the thermal conductivity of the sample made from nano-powder was about 20% higher than that of the one made from micro-powder (fig. 3 (b)). The two samples had similar electrical conductivities. In other words, the two samples had essentially the same thermal conductivity attributed to electrons. There are possibly two reasons for this phenomenon. One is the presence of MgO in the sintered body. The thermal conductivity of MgO is higher than that of Mg₂Si. Even a small amount of MgO at the grain boundary of Mg₂Si would raise the thermal conductivity of the sample. The second reason is high bulk density. It has been shown already that the relative density of the bulk sample made from nano-powder (99.8%) was higher than that of the sample made from micro-powder (95.4%). There were a few micro pores in the lower density bulk sample. It is thought that these micro pores decreased the thermal conductivity. Figure 4 shows the temperature dependence of the dimensionless figure of merit ($ZT$) of the Mg₂Si bulk sample made from micro-powder (~75 μm) [18] and the sample made from nano-powder. $ZT$ of the sample made from nano-powder was lower than that of the sample made from micro-powder, because the thermal conductivity of the nano-powder sample was higher in the temperature range from 323 K to 823 K.

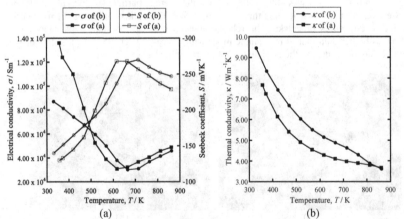

(a)                                                                 (b)

Figure 3. Temperature-dependent Seebeck coefficient ($S$), electrical conductivity ($\sigma$), and thermal conductivity ($\kappa$) of Mg₂Si bulk made from micro-powder (~75 μm) [17] (a) and Mg₂Si bulk made from nano-powder (b).

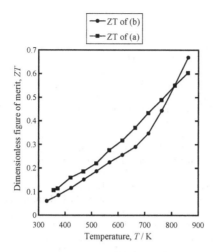

Figure 4. Temperature-dependent dimensionless figure of merit (*ZT*) of Mg$_2$Si bulk made from micro-powder (~75 μm) [18] (a) and Mg$_2$Si bulk made from nano-powder (b).

## CONCLUSIONS

We succeeded in fabricating Mg$_2$Si nano-powder by dry ball milling. The powder had particle sizes from about 100 nm to 3 μm. The most common particle size was in the range of 400 nm ~ 600 nm, and the average particle size was about 500 nm. The spark plasma sintering method produced high-density Mg$_2$Si bulk from these nano-powders. The maximum *ZT* of the Mg$_2$Si bulk was 0.67 at 873 K. On the other hand, the fabricated Mg$_2$Si bulk contained a small amount of MgO, which has high thermal conductivity. As a result, the thermal conductivity of the Mg$_2$Si bulk from nano-powder was higher than that of the sample made from micro-powder. It can be expected that reducing the oxidation of Mg$_2$Si during these processes will increase *ZT*.

## REFERENCES

1. M. W. Heller and G. C. Denielson, *J. Phys. Chem. Solids* **23**, 601 (1962)
2. R. J. LaBotz, D. R. Mason, and D. F. O'Kane, *J. Electrochem. Soc.* **110**, 121 (1963)
3. Y. Noda, H. Kon, Y. Furukawa, N. Otsuka, I. Nishida, and K. Matsumoto, *Mater. Trans. JIM* **33**, 845 (1992)
4. J. Tani and H. kido, *Physica B* **223**, 364 (2005)
5. V. K. Zaitsev, M. I. Fedorv, E. A. Gurieva, I. S. Eremin, P. P. Konstantinov, A. Yu. Samunin, and M. V. Vedernikov, Phys. Rev. **B74**, 045207 (2006)
6. M. Fukumoto, T. Iida, K. Makino, M.Akasaka, Y. Oguni and Y. Takanashi, *Mater. Res. Soc. Proc.* **1004**, U6.13.1-U6.13.6 (2008)

7. Y. Oguni, T. Iida, A. Matsumoto, T. Nemoto, J. Onosakal, H. Takaniwa, T. Sakamoto, D. Moril, M. Akasaka, J. Sato, T. Nakajima, K. Nishio, and Y. Takanashi, *Mater. Res. Soc. Symp. Proc.***1044**, pp. 413 (2008)
8. T. Nemoto, T. Iida, Y. Oguni, J. Sato, A. Matsumoto, T. Sakamoto, T. Miyata, T. Nakajima, H. Taguchi, K. Nishio and Y. Takanashi, *Mater. Res. Soc. Symp. Proc.* **1166**, pp.141 (2009)
9. T. Nemoto, T. Iida, J. Sato, Y. Oguni, A. Matsumoto, Y. Miyata, T. Sakamoto, T. Nakajima, H. Taguchi, K. Nishio, and Y. Takanashi, *J. of Electronic Mater.*, **39**, 9, 1572-1578 (2010)
10. K. Arai, H. Akimoto, T. kineri, T. Iida, K. Nishio, *Key Eng. Mater.*, **485**, Electroceramics in Japan XIV, 169-172 (2011)
11. T. Sakamoto, T.Iida, Y. Taguchi, S. Kurosaki, Y. Hayatsu, K. Nishio, Y.Kogo, and Y. Takanashi, *J. of Electronic Mater.*, **41**, Issue 6, 1429-1435 (2012)
12. K. Arai, M. Matsubara, Y. Sawada, T. Sakamoto, T. Kineri, Y. Kogo, T. Iida, and K. Nishio, *J. of Electronic Mater.*, **41**, Issue 6, 1771-1777 (2012)
13. B. Poudell, Q. Hao, Y. Ma, Y. Lan, A. Minnich, B. Yu, X. Yan, D. Wang, A. Muto, D. Vashaee, X. Chen, J. Liu, M. S. Dresselhaus, G. Chen, and Z. Ren, *Science*, **320**, 634-638 (2008)
14. G. Joshi, H. Lee, Y. Lan, X. Wang, G. Zhu, D. Wang, R. W. Gould, D. C. Cuff, M. Y. Tang, M. S. Dresselhaus, G. Chen and Z. Ren, *Nano Lett.*, **8** (12), 4670-4674 (2008)
15. Y. W. Gu, K. A. Khor, P. Cheang: *Biomater.*, **25**, 4127 (2004)
16. D. Wang, L. Chen, Q. Yao, J. Li: *Solid State Com.*, 615 (2004)
17. JCPDS No.34-0458
18. T. Sakamoto, T. Iida, N. Fukushima, Y. Honda, M. Tada, Y. Taguchi, Y. Mito, H. Taguchi, Y. Takanashi, *Thin Solid Films*, **519**, Issue 24. 8528-8531 (2011)

Mater. Res. Soc. Symp. Proc. Vol. 1490 © 2012 Materials Research Society
DOI: 10.1557/opl.2012.1571

# Thermoelectric properties of Li-doped $Cu_{0.95-x}M_{0.05}Li_xO$ (M=Mn, Ni, Zn)

N. Yoshida, T. Naito and H. Fujishiro*
(* corresponding author: fujishiro@iwate-u.ac.jp)
Faculty of Engineering, Iwate University, 4-3-5 Ueda, Morioka 020-8551, Japan.

ABSTRACT

Thermoelectric properties of the Li-doped $Cu_{0.95-x}M_{0.05}Li_xO$ (M=divalent metal ion; Mn, Ni, Zn) were investigated at the temperature up to 1273 K. In the doped divalent metal ions, $Zn^{2+}$ ion was the most effective to reduce the thermal conductivity, and the $Ni^{2+}$ substitution was preferable to decrease the electrical resistivity. For the $Cu_{0.95-x}Ni_{0.05}Li_xO$ sample at $x$=0.03, the maxima of the dimensionless thermoelectric figure of merit $ZT$ and the power factor $P$ at 1246 K were $4.2 \times 10^{-2}$ and $1.6 \times 10^{-4}$ W/K$^2$m, respectively. The enhancement of the thermoelectric properties of the Li-doped $Cu_{0.95-x}M_{0.05}Li_xO$ system was discussed.

INTRODUCTION

The thermoelectric technology using a waste heat for electric power generation devices has been revised because of recent energy crisis besides other renewable energy sources. Since the discovery of $NaCo_2O_4$ in 1998 [1], oxide thermoelectric materials have been intensively studied using Ca-Co-O and $NaCo_2O_4$ systems with layered Co-O octahedron [2,3], as a substitute for the conventional semiconductors such as $Bi_2Te_3$. However, the thermoelectric performance has been still low. In material researches for thermoelectricity, we focus on the possibility of a conventional CuO system, because it is inexpensive, plentiful and non-toxic resources. CuO is known to be a metal deficient $p$-type semiconductor with a band gap of 1.2 eV with monoclinic crystal structure, and many studies were reported for the physical properties [4,5]. High-purity CuO has a large positive Seebeck coefficient $S$, but shows both a high electrical resistivity $\rho$ and a high thermal conductivity $\kappa$. If the doping or substitution of univalent or trivalent ions is possible for the Cu site, the electrical resistivity and the thermal conductivity might be decreased. In this case, the thermoelectric efficiency may be enhanced, if the Seebeck coefficient is not deteriorated so much. Similarly to the CuO system, ZnO is a plentiful and promising material for the $n$-type thermoelectricity with a simple wurtzite structure [6]. The electrical resistivity $\rho$ drastically decreases by the doping of a small amount of $Al^{3+}$ and/or $Ga^{3+}$, and the dimensionless thermoelectric figure of merit $ZT$ ($=S^2T/\rho\kappa$) reaches 0.3 and 0.65 at 1273 K for $Zn_{0.98}Al_{0.02}O$ and $Zn_{0.96}Al_{0.02}Ga_{0.02}O$, respectively [7,8]. For the CuO system, the electric conductivity $\sigma$ ($=1/\rho$) and Seebeck coefficient $S$ were reported for the Li and Al doping, in which the Li-doping enhances the $p$-type electrical conductivity [9]. However, a systematic investigation has not been performed as a thermoelectric material of CuO as for the doping species and the optimum doping concentration.

Recently, we have reported the alkali metal substitution for the Cu-site in CuO, in which $Li^+$ is the most promising ion to enhance the thermoelectric performance. The $ZT$ takes a maximum of $3.2 \times 10^{-2}$ at 1246 K for the $Cu_{0.97}Li_{0.03}O$ [10]. To enhance the thermoelectric performance, the reduction of the thermal conductivity is a possible approach, which is usually realized by the substitution of the element with different ionic radius.

In this study, we fabricated the Li-doped $Cu_{0.95-x}M_{0.05}Li_xO$ (M=Mn, Ni, Zn) and measured thermoelectric properties at high temperature. The potential of the CuO system as thermoelectric materials was discussed.

## EXPERIMENT

$Cu_{0.95-x}M_{0.05}Li_xO$ polycrystals doped with a divalent-metal element (M=Mn, Ni, Zn) were prepared by a conventional solid-state reaction. The raw powders of CuO (Furuuchi Chemical; 99.9%), MnO (KOJUNDO; 99.9%), NiO (Furuuchi Chemical; 99.9%), ZnO (KOJUNDO; 99.9%) and $Li_2O$ (KOJUNDO; 99% up) were mixed with a molar ratio of Cu:M:Li=0.95-$x$:0.05:$x$ ($x$=0.01, 0.03). The specimen was pressed cold-isostatically under 64 MPa into a bar and then sintered at 1060°C for 10 h in oxygen flow. The measured densities of each sample were greater than 95% of the ideal density. The rectangular-shaped specimen with about 2 x 2 x 10 $mm^3$ was cut from the bar for the measurements of the thermoelectricity. Powder X-ray diffraction measurements were performed at 300 K using a Cu $K\alpha$ radiation (Rigaku; Multi Flex) in the range of $20° \leqq 2\theta \leqq 90°$ with 0.02° steps. The electrical resistivity $\rho(T)$ and Seebeck coefficient $S(T)$ were simultaneously measured in the temperature range from 300 to 1273 K using an automated measuring system (Ozawa Science; RZ2001i) and the thermoelectric power factor $P=S^2/\rho$ was calculated. Low-temperature thermal conductivity $\kappa(T)$ was measured by a steady-state heat flow method from 10 to 300 K using a home-made apparatus combined with helium refrigerator [11]. High-temperature thermal conductivity $\kappa(T)$ was also measured by a laser flash method (Ulvac-Riko; TC-7000) up to 1023 K. $\kappa(T)$ was extrapolated up to 1300 K using a fitting curve which was proportional to $T^{-1}$ for the experimental $\kappa(T)$ results at low and high temperatures. A dimensionless figure of merit $ZT=S^2T/\rho\kappa$ was estimated using these values.

## DISCUSSION

### Thermal conductivity of $Cu_{0.95}M_{0.05}O$ (M=Mn, Ni, Zn)

The doping effect of divalent metal (M) ion in CuO on the thermal conductivity $\kappa(T)$ was investigated. Figure 1 shows the temperature dependence of $\kappa(T)$ of pure CuO and $Cu_{0.95}M_{0.05}O$ (M=Mn, Ni, Zn). $\kappa(T)$ was reduced by the substitution for the Cu-site, in which the Zn substitution is the most effective. The $\kappa(T)$ reduction by the divalent metal ion doping seems to be the introduction of the lattice disorder due to the substitution of the M ion for the Cu site with difference of the ionic radius. $\kappa(T)$ of materials can be generally represented by the sum of lattice contribution ($\kappa_{ph}$) and carrier contribution ($\kappa_e$). In the present samples, heat propagates almost due to phonons ($\kappa_{ph}$) because of the high $\rho(T)$ value. The measured $\kappa(T)$ shows a broad peak at low temperature, which results from the phonon-phonon scattering as can be seen in insulating materials. In the previous paper, we reported that the thermoelectric properties were enhanced by the Li doping to the CuO matrix [10]. In the later subsection, we report on the Li doping effect on the thermoelectric properties of $Cu_{0.95}M_{0.05}O$ with smaller $\kappa(T)$.

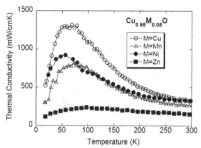

**Figure 1.** Temperature dependence of the thermal conductivity $\kappa(T)$ of pure CuO and $Cu_{0.95}M_{0.05}O$ (M=Mn, Ni, Zn).

## Thermoelectric properties of Li-doped $Cu_{0.95-x}M_{0.05}Li_xO$ (M=Ni, Zn)

Figures 2(a) and 2(b) show the temperature dependence of the electrical resistivity $\rho(T)$ and Seebeck coefficient $S(T)$ for the Li-doped $Cu_{0.95-x}M_{0.05}Li_xO$ (M=Ni, Zn). In Fig. 2(a), $\rho(T)$ of the $Cu_{0.95}M_{0.05}O$ samples decreased with increasing $T$ at $T<800$ K, compared with pure CuO, but was comparable with that of the pure CuO at higher temperatures. Similarly to the Li-doped CuO [10], the Li doping can drastically reduce the $\rho(T)$ and the absolute value of $\rho(T)$ decreased with increasing contents of Li. $\rho(T)$ of the $Cu_{0.95-x}Ni_{0.05}Li_xO$ is lower than that of the $Cu_{0.95-x}Zn_{0.05}Li_xO$ at identical $x$, both of which are lower than that of the Li-doped $Cu_{1-x}Li_xO$. In Fig. 2(b), $S(T)$ decreases with increasing $x$. $S(T)$ of the $Cu_{0.95-x}Ni_{0.05}Li_xO$ is lower than that of the $Cu_{0.95-x}Zn_{0.05}Li_xO$ at identical $x$, both of which are lower than that of the Li-doped $Cu_{1-x}Li_xO$ [10]. Because $S(T)$ is closely correlated with $\rho(T)$ [12].

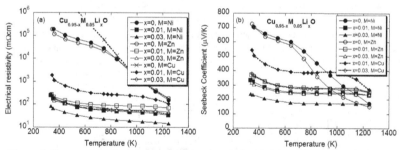

**Figure 2.** Temperature dependence of (a) the electrical resistivity $\rho(T)$ and (b) the Seebeck coefficient $S(T)$ of pure CuO and Li-doped $Cu_{0.95-x}M_{0.05}Li_xO$ (M=Ni, Zn).

Figure 3(a) shows the $x$ dependence of the thermal conductivity $\kappa(T)$ of the Li-doped $Cu_{0.95-x}M_{0.05}Li_xO$ (M=Ni, Zn). As shown in Fig. 1, the $\kappa(T)$ reduction by the Zn substitution was larger than that by the Ni substitution, both of which $\kappa(T)$ further decreased with increasing

71

content of Li, $x$. The origin of the additional $\kappa(T)$ reduction by the Li doping comes from the introduction of the disorder in the lattice.

Figure 3(b) shows the $\kappa(T)$ of the $Cu_{0.94}Ni_{0.05}Li_{0.01}O$ sample up to 1000 K. The fitting curve, which was proportional to $T^{-1}$, was also shown and was extrapolated up to 1273 K. The estimated $\kappa(T)$ value was 52 mW/cmK at 1200 K, which was smaller than that for the $Cu_{0.99}Li_{0.01}O$ sample [10] due to the enhanced disorder by the additional Ni substitution.

Figure 4(a) depicts the temperature dependence of the thermoelectric power factor $P(T)$ of the Li-doped $Cu_{0.95-x}M_{0.05}Li_xO$ (M=Ni, Zn). $P(T)$ of the $Cu_{0.97}Li_{0.03}O$, which showed the highest $P(T)$ in the previous paper [10] is also shown. As would be expected from the results of $\rho(T)$ and $S(T)$ values, $Cu_{0.95-x}Ni_{0.05}Li_xO$ ($x$=0.03) shows the highest $P(T)$ of $2.0 \times 10^{-4}$ W/K²m at 1246 K, which is higher than that for $Cu_{0.97}Li_{0.03}O$.

Figure 4(b) shows the dimensionless figure of merit $ZT$ of the Li-doped $Cu_{0.95-x}M_{0.05}Li_xO$ (M=Ni, Zn) as a function of $T$. The $ZT$ value increased with increasing temperature and contents of Li, and took a maximum value of $4.2 \times 10^{-2}$ for $Cu_{0.95-x}Ni_{0.05}Li_xO$ ($x$=0.03) at 1246 K, which was about 30% larger than that of $Cu_{0.97}Li_{0.03}O$. The enhancement of the $ZT$ value was also confirmed for the $Cu_{0.95-x}Zn_{0.05}Li_xO$ system. The reduction of $\kappa(T)$ and $\rho(T)$ by the Ni or Zn substitution besides the Li doping enhances the thermoelectric properties.

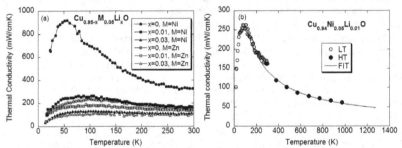

**Figure 3.** Temperature dependence of the thermal conductivity $\kappa(T)$ of $Cu_{0.95-x}M_{0.05}Li_xO$ (M=Ni, Zn) at $T{\leq}300$ K and (b) $\kappa(T)$ for $Cu_{0.94}Ni_{0.05}Li_{0.01}O$ up to 1000 K. The extrapolated fitting curve up to 1300 K is also shown.

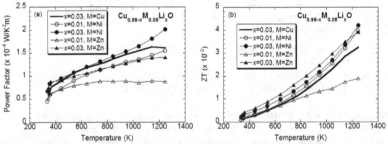

**Figure 4.** Temperature dependence of the (a) power factor $P$ and (b) dimensionless figure of merit $ZT$ of $Cu_{0.95-x}M_{0.05}Li_xO$ (M=Ni, Zn).

## CONCLUSIONS

We have investigated the thermoelectric properties of the Li-doped $Cu_{0.95-x}M_{0.05}Li_xO$ (M=Mn, Ni, Zn) as thermoelectric materials. Important experimental results and conclusions obtained from this study are summarized as follows.
1) Among the divalent metal substitution for the Cu-site in $Cu_{0.95}M_{0.05}O$, $Zn^{2+}$ is the most promising ions to decrease the thermal conductivity. The $\kappa(T)$ reduction seems to be the effect of the lattice disorder by the substitution of Zn for the Cu site.
2) For the $Cu_{0.95-x}M_{0.05}Li_xO$ (M=Ni, Zn), the electrical resistivity $\rho(T)$, Seebeck coefficient $S(T)$ and the thermal conductivity $\kappa(T)$ decreased with increasing contents of Li up to $x$=0.03. Power factor $P(T)=S^2/\rho$ increased with increasing contents of Li, takes a maximum of $P$=2.0×10$^{-4}$ W/K$^2$m at 1246 K.
3) For the $Cu_{0.92}Ni_{0.05}Li_{0.03}O$, the dimensionless thermoelectric figure of merit $ZT=S^2T/\rho\kappa$ increased with increasing contents of Li. A maximum of $ZT$=4.2×10$^{-2}$ was realized at 1246 K which was about 30% larger than that of $Cu_{0.97}Li_{0.03}O$.

## ACKNOWLEDGMENTS

The authors thank Prof. Takashi Goto and Prof. Rong Tu of Tohoku University, Japan, for the assistance of thermal conductivity measurement using Laser flash method and valuable discussion. This work was performed under the Inter-university Cooperative Research Program of the Institute for Materials Research, Tohoku University. The presentation at the conference was financially supported by Iwate University.

## REFERENCES

1. I. Terasaki, Y, Sasago, and K. Uchinokura: Phys. Rev. B **56**, R12685 (1997).
2. R. Funahashi, I. Matsubara, H. Ikuta, T. Takeuchi, U. Mizutani, and S. Sodeoka: Jpn. J. Appl. Phys. **39**, L1127 (2000).
3. M. Ito, T. Nagira, D. Furumoto, S. Katsuyama, and H. Nagai: Scripta Materialia **48**, 403 (2003).
4. Y. K. Jeong and G. M. Choi: J. Phys. Chem. Solids **57**, 81 (1996).
5. B. Yang, T. Thurston, J. Tranquada, and G. Shirane: Phys. Rev. B **39**, 4343 (1989).
6. D. Klimm, S. Ganschow, D. Schulz, and R. Fornari: J. Cryst. Growth **310**, 3009 (2008).
7. T. Tsubota, M. Ohtaki, K. Eguchi, and H. Arai: J. Mater. Chem. **7**, 85 (1997).
8. M. Ohtaki, K. Araki, and K. Yamamoto: J Electron. Mater. **38**, 1234 (2009).
9. S. Suda, S. Fujitsu, K. Koumoto and H. Yanagida: Jpn. J. Appl. Phys. **31**, 2488 (1992).
10. N. Yoshida, T. Naito and H. Fujishiro: submitted to Jpn. J. Appl. Phys. (2012).
11. H. Fujishiro, M. Ikebe, T. Naito, K. Noto, S. Kobayashi, and S. Yoshizawa: Jpn. J. Appl. Phys. **33**, 4965 (1994).
12. J. M. Ziman: *PRINCIPLES OF THE THEORY OF SOLIDS* (CAMBRIDGE UNIVERSITY PRESS, London, 1964) p. 200.

Mater. Res. Soc. Symp. Proc. Vol. 1490 © 2013 Materials Research Society
DOI: 10.1557/opl.2013.150

# Investigation of the valence band structure of PbSe by optical and transport measurement

Thomas C. Chasapis[1], Yeseul Lee[1], Georgios S. Polymeris[2], Eleni C. Stefanaki[2], Euripides Hatzikraniotis[2], Xiaoyuan Zhou[3], Ctirad Uher[3], Konstantinos M. Paraskevopoulos[2] and Mercouri G. Kanatzidis[1]
[1]Department of Chemistry, Northwestern University, Evanston, Illinois, 60208, U.S.A.
[2]Physics Department, Aristotle University of Thessaloniki, GR- 54124, Thessaloniki, Greece
[3]Department of Physics, University of Michigan, Ann Arbor, Michigan 48109, U.S.A.

## ABSTRACT

We investigated the valence band structure of PbSe by a combined study of the optical and transport properties of $p$-type $Pb_{1-x}Na_xSe$, with Na concentrations ranging from $0 - 4\%$, yielding carrier densities in a wide range of $10^{18} - 10^{20}$ $cm^{-3}$. Room temperature infrared reflectivity studies showed that the susceptibility (or conductivity) effective mass $m^*$ increases from ~ $0.06m_0$ to ~ $0.5m_0$ on increasing Na content from $0.08\%$ to $3\%$. The Seebeck coefficient scales with doping in the whole temperature range, yielding lower values for higher Na contents, while the Hall coefficient increases on heating from room temperature showing a peak close to 650 K. The room temperature Pisarenko plot is well described by the simple parabolic band model up to ~ $1 \cdot 10^{20}$ $cm^{-3}$. In order to describe the behaviour in the whole concentration range, the application of the two band model, i.e. light hole and heavy hole, was used giving density of states effective masses $0.28m_0$ and $2.5m_0$ for the two bands respectively.

## INTRODUCTION

The lead chalcogenides, such as PbTe and PbSe have been under consideration for several decades as semiconductors of potential interest in the field of thermoelectrics. Between them PbSe has several advantages with respect to thermoelectric application since Se is much more abundant than Te and has lower thermal conductivity, which is beneficial for a larger figure of merit [1-4].

Recent calculations suggest that $p$-type doped PbSe with carrier densities $1.2 - 1.7 \cdot 10^{20}$ $cm^{-3}$ may reach $ZT \sim 2$ at temperatures near 1000 K due to the appearance of a flat, high mass, high DOS band of ~ 0.35 eV (at 0 K) below the $L$ point valence band edge, which enhances the thermopower [5]. Using first-principles approach Peng $et$ $al.$ [1] explored the band structure of Na doped PbSe demonstrating also the existence of flat bands that lead to increased DOS and larger Seebeck coefficients relative to the pristine material. Wang $et$ al. studied the thermoelectric properties of $Na_xPb_{1-x}Se$ system with carrier densities up to $2.6 \cdot 10^{20}$ $cm^{-3}$ and found that their room temperature Seebeck data versus hole carrier density (i.e. the Pisarenko plot) may be explained well by a single parabolic model [6]. The deviations observed at higher temperatures were attributed to the contribution of a second heavy hole band. The application of both a single non-parabolic band model and a two-band model in the case of the room temperature Pisarenko plot of the $K_xPb_{1-x}Se$ system with hole densities up to ~ $1.5 \cdot 10^{20}$ $cm^{-3}$ showed that most of the contribution comes from the non-parabolic $L$ band [7].

All the above suggest that the picture concerning the valence band structure of $p$-type PbSe is still ambiguous for whether or not there is a contribution of a second hole band regarding the transport properties. In the present study probed this question through a combined study of optical and transport properties of heavily doped $p$-type $Na_xPb_{1-x}Se$, with densities up to $\sim$ $6.2 \cdot 10^{20}$ cm$^{-3}$. We show that a two-band model is more consistent with both optical and transport investigations.

## EXPERIMENT

Ingots ($\sim$10 g) with nominal compositions $Pb_{1-x}Na_xSe$ (x = 0 – 0.04) were prepared by mixing appropriate ratios of elemental Pb (99.99%, American Elements), Se (99.999%, 5N plus Inc), and Na (99.95%, Aldrich) in quartz tubes with carbon coating under a $N_2$-filled glove box. The tubes were sealed under vacuum ($\sim$10$^{-4}$ Torr) and heated up to 1150 °C over a period of 12 h, soaked at that temperature for 5 h, slowly cooled to 1080 °C in 12 h, and rapidly cooled to room temperature over 3 h.

Room-temperature infrared reflectivity (IR) measurements were performed on finely polished PbSe samples using a Bruker 113V FTIR spectrometer with the standard circular iris of 8 mm in diameter. The spectra were collected in the 400–4000 cm$^{-1}$ spectral region with a resolution of 2 cm$^{-1}$ at nearly normal incidence. The reflection coefficient was determined by a typical sample-in–sample-out method with a mirror as the reference. The samples for charge transport properties measurement were cut and polished into a parallelepiped with dimensions of ~2 mm × 3 mm × 10 mm. The Seebeck coefficient $S$ was measured under a helium atmosphere

($\sim$ 0.1 atm) from room temperature to $\sim$ 700 K using a ULVAC-RIKO ZEM-3 system. Carrier

concentrations, $p_h$, were determined using measurements of Hall coefficients at room temperature with a home-build system in applied magnetic field. The high temperature Hall coefficient was measured in a home-made high temperature apparatus. The Hall resistance was monitored with a Linear AC Resistance Bridge (LR 700), and the data were taken in a field ±1.25 T.

## DISCUSSION

### Infrared Reflectivity

Room temperature reflectivity spectra measured for a selection of samples with different Na content are presented in figure 1(a). It is readily seen that the reflectivity minimum, associated with the free carriers' collective oscillations, namely the plasma frequency $\omega_P$, shifts towards higher frequencies for increasing doping level. Another experimental observation deals with the behavior of the reflectivity level at large wavenumbers which in turn decreases with increasing Na concentration. The spectra were analysed by means of the Drude model for the complex dielectric function [8]. In the case of the Drude model, the plasma frequency $\omega_P$ is related to the the carrier concentration $N$ via the following equation:

$$\omega_p^2 = \frac{N \cdot e^2}{\varepsilon_0 m^* \varepsilon_\infty} \tag{1}$$

where $\varepsilon_\infty$ is the optical dielectric constant, $\varepsilon_0$ is the vacuum permeability, $m^*$ is the free carriers conductivity effective mass and $e$ is the electron charge. The results of the analysis are summarized in figures 1((a)-(c)).

Starting with Eq. (1), the increase of the hole concentration is giving rise to increased $\omega_P$ values, as can be seen from figure 1(b), which is actually reflecting the blue-shift of the reflectivity minimum of figure 1(a). The remaining fitting obtained parameters, i.e. the $\varepsilon_\infty$ and the $m^*$ are strongly related to the band structure, making infrared studies a powerful tool in the field of thermoelectrics concerning the band structure. As can be seen from the inset of figure 1(a) the optical dielectric constant, which is actually related to the high frequency reflectivity level, is monotonically decreased with the increase of the hole concentration. Considering the fact that $\varepsilon_\infty$ is inversely proportional to the band gap [9] this behavior may be attributed to a widening of the optically extracted gap due to the increase of the Fermi level on doping, known as Burstein-Moss shift [10].

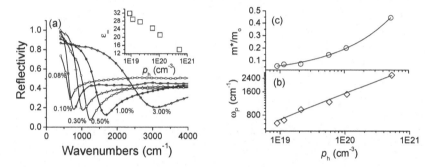

**Figure 1:** (a) Room-temperature reflectivity spectra for ingots samples with various Na contents. Inset: The dependence of the high-frequency dielectric constant on carrier concentration $p_h$. (b) The plasma frequency, reflecting the reflectivity minimum, is blue-shifted from on going from 0.08% to 3% Na. (c) The conductivity effective mass is strongly enhanced for $p_h \geq 6 \cdot 10^{19}$ cm$^{-3}$ reaching the value of $0.5 m_0$ for the 3% Na composition.

The hole effective mass $m^*$ was found to increase with increasing hole density and the increase was more pronounced for the compositions with hole concentrations greater than $\sim 6 \cdot 10^{19}$ cm$^{-3}$. In general, the continuous variation of the $m^*$ indicates that either the energy band involved is non-parabolic or the valence band feature can be modeled using multiple offset valence bands. Veis et al. found that the effective masses of holes in PbSe increased with the hole concentration more rapidly than predicted by the non-parabolic Kane model and they interpreted their results using a two-band model. The existence of underlying bands was confirmed by the room temperature absorption coefficient spectra. According to this interpretation there is a heavy hole band located 0.23 eV lower than the $L$ point light band, and it was found to be responsible for the increase of the conductivity effective mass on increasing hole density [11]. According to our results one may say that the observed increase of $m^*$ in the higher hole density region is consistent with evidence for a heavy hole band contribution, although further studies are needed.

## Transport Measurements

The Seebeck coefficient, $S$, of the Na-doped PbSe specimens as a function of temperature is shown in figure 2 (a). For all the studied compositions $S$ increases monotonically on increasing temperature from 300 K to 700 K, and scales with Na content in the whole temperature range. The room temperature values range from ~ 100 µV/K to ~ 26 µV/K when the Na content increases from 0.10% to 4.00%. The temperature dependent Hall coefficient is shown in figure 2(b), giving gradually increased values and a peak close to 650 K.

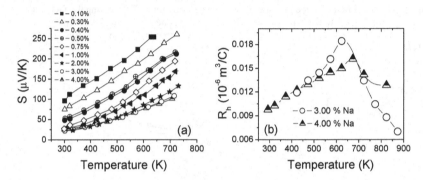

**Figure 2.** (a) Seebeck coefficient, $S$, as a function of temperature for $Pb_{1-x}Na_xSe$ compositions. (b) Hall coefficient, $R_h$, as a function of temperature for two selected samples. In each case the $R_h(T)$ is peaked at ~ 650 K.

Important information about the band structure can be revealed with a plot of $S$ versus $p_h$ (the Pisarenko plot) at a constant temperature [12]. Figure 3 shows the Pisarenko plot at room temperature and the experimental data fall satisfactorily on a curve based on single parabolic model (short dashed line), with $m = 0.29m_0$ and ing acoustic phonon scattering. This value is in good agreement with the $0.28m_0$ obtained by Wang et al. [6]. As it obvious from figure 3, the agreement between the theoretical line and the experimental data is better for hole concentrations up to ~ $1 \cdot 10^{20}$ cm$^{-3}$. For higher carrier concentrations the Pisarenko flattens and the single parabolic model falls well below the measured data.

The deviation of $S$ on $p_h$ from the trend expected from the single parabolic model at high carrier concentrations has also been observed in the case of the $K_xPb_{1-x}Te$ [7] and in $Na_{1-x}Pb_xTe$ [13] for which the existence and significant contribution of a heavy hole band is well known [14]. This behavior has been explained using a two band model with non- parabolic light hole $L$ band, described by a Kane model and a parabolic heavy hole $\Sigma$ band, with room temperature energy difference between the two bands, $\Delta E_v$, ~ 0.06 eV [7].

As already mentioned, the two band model with $\Delta E_v = 0.23$ eV describes well the dependence of the room temperature conductivity effective mass on hole concentration in the case of $p$-type PbSe [11]. Also the maximum value of the Hall coefficient of figure 2(b), may be qualitatively explained on a two band Hall coefficient model as a function of temperature. According to Allgaier, the position of the $R_h(T)$ maximum is specified by the energy difference $\Delta E_v$ [14]. In our case and for the 3% and 4% Na-doped compositions, the maximum occurs in the region of ~ 650 K (see figure 2(b)), while for the PbTe codoped with 1.25% K and with 0.8% Na the $R_h(T)$ maximum was found at ~ 415 K [15]. In the framework of the two-band model, the changing position of the $R_h(T)$ maximum from PbSe to PbTe, may be attributed to different $\Delta E_v$ values [14].

Based on the above observations we employed the two band model to describe the $S$ vs $p_h$ of figure 3 in the whole concentration range. For energy difference $\Delta E_v$ we used the optically determined room temperature value 0.23 eV [11], while the $L$ band non-parabolicity parameter was determined by the temperature dependence of the band gap [16].

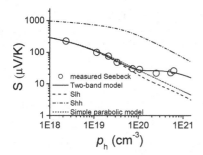

$p_h$ (cm$^{-3}$)

**Figure 3:** Room temperature Pisarenko plot of the Na$_x$Pb$_{1-x}$Se system. The single parabolic model (short dashed line) with $m = 0.29m_o$ seems to describe well the experimental data up to ~ $1 \cdot 10^{20}$ cm$^{-3}$. The application of the two-band model (solid line) describes the $S$ vs $p_h$ data fairly good in the whole concentration range. The two contributions, i.e. from the light hole $S_{lh}$ (dashed line) and from the heavy hole $S_{hh}$ (dash-dotted line) are also shown. The contribution of the $S_{hh}$ is clear for hole densities > $1 \cdot 10^{20}$ cm$^{-3}$.

The band masses, $m_{lh}$ and $m_{hh}$, and the carrier mobilities ratio, $\mu_{lh}/\mu_{hh}$, were the fitting adjustable parameters between the measured and the calculated Seebeck coefficients. The best fit calculated curve, shown as solid line in figure 3, was achieved for $m_{lh} = 0.28m_o$ and $m_{hh} = 2.5m_o$ in good agreement with literature [7, 11] and $\mu_{lh}/\mu_{hh} = 6.2$. We must notice that for simplicity we kept the mobility ratios independent of the hole density, although there might be a concentration dependence. The Seebeck coefficients of the two bands, i.e. the light and the heavy mass bands, are also shown in figure 3. It is obvious that the non-parabolic light hole band seems to mostly contribute up to ~ $7 \cdot 10^{19}$ cm$^{-3}$ and it looks to describe quite well the measured data, in accordance with the observations of Zhang *et al.* for the K$_x$Pb$_{1-x}$Se system [7]. We must notice here that $S_{lh}$ actually lies close to the respective line of the single parabolic model. Thus we may conclude that for carrier concentrations up to ~ $1 \cdot 10^{20}$ cm$^{-3}$ both models describe well the Seebeck concentration dependence. However going to higher hole densities, the situation is more apparent and the onset of the Seebeck flattening originates from the heavy hole band contribution.

## CONCLUSIONS

In this work we studied the optical and transport properties of the $p$-type $Pb_{1-x}Na_xSe$ system, with $0 \leq x \leq 0.04$. The free carrier conductivity effective masses extracted from the analysis of the experimental reflectivity spectra were found to increase on increasing Na content. Based on heavily doped composition the dependence of the room temperature Seebeck coefficient on carrier concentration can be described in the whole concentration range by the incorporation of a second band located 0.23 eV lower from the valence band edge with the density of states effective mass being $2.5m_0$. The two band model appears to be consistent in accounting for both the optical and transport properties.

## ACKNOWLEDGMENTS

This material is based upon work supported as part of the Revolutionary Materials for Solid State Energy Conversion, an Energy Frontier Research Center funded by the U.S. Department of Energy, Office of Science, Office of Basic Energy Sciences, under Award No. ED-SC 0001054.

## REFERENCES

1. H. Peng, J-H. Song, M.G. Kanatzidis and A.J. Freeman, *Phys. Rev. B*, **84**, 125207 (2010).
2. J. Androulakis, D.Y. Chung, X. Su and L. Zhang, C. Uher, T.C. Hasapis, E. Hatzikraniotis, and K.M. Paraskevopoulos, M.G. Kanatzidis, *Phys. Rev. B*, **84**, 155207 (2011).
3. J. Androulakis, I. Todorov, JQ. He, D.Y. Chung, V. Dravid, M.G. Kanatzidis, *J. Am. Chem. Soc.*, **133**, 10920 (2011).
4. J. Androulakis, Y. Lee, I. Todorov, D.Y. Chung, M.G. Kanatzidis, *Phys. Rev. B*, **83**, 195209 (2011).
5. D. Parker and D.J. Singh, *Phys. Rev. B*, **82**, 035204 (2010).
6. H. Wang, Y. Pei, AD. LaLonde and G.J. Snyder, *Adv. Mater.*, **23**, 1366 (2011).
7. Q. Zhang, F. Cao, W. Liu, K. Lukas, Bo Yu, S. Chen, C.Opeil, D. Broido, G. Chen and Z. Ren, *J. Am. Chem. Soc.*, **134**, 10031 (2012).
8. A. A. Kukharskii, *Solid State Commun.* **13**, 1761 (1973).
9. R.N. Tauber and I.B. Cadoff, *J. Appl. Phys.*, **38**, 3714 (1967).
10. I. Hamberg and C. G. Granqvist, K.-F. Berggren, B.E. Sernelius, and L. Engstrom, *Phys. Rev. B*, **30**, 3240 (1984).
11. A.N. Veis, R.F. Kuteinikov, S.A. Kumzerov, and Yu. I. Ukhanov, *Sov. Phys. Semicond.*, **10**, 1320 (1976).
12. J.P. Heremans, B. Wiendlocha and A. M. Chamoire, *Energy Environ. Sci.*, **5**, 5510 (2012).
13. Y. Pei, X. Shi, A. LaLonde, H. Wang, L. Chen and G.J. Snyder, *Nature*, **473**, 66 (2011).
14. R. S. Allgaier, *J. Appl. Phys.*, **36**, 2429 (1965).
15. J. Androulakis, I. Todorov and D.Y. Chung, S. Ballikaya, G. Wang and C. Uher, M.G. Kanatzidis, *Phys. Rev. B.*, **82**, 115209 (2010).

16. Yu.I. Ravich, B.A. Efimova and I.A Smirnov, "Semiconducting Lead Chalcogenides", ed. L.S. Still'bans (Plenum Press, 1970) p. 157.

Mater. Res. Soc. Symp. Proc. Vol. 1490 © 2013 Materials Research Society
DOI: 10.1557/opl.2013.25

# Optimizing Thermoelectric Efficiency of La$_{3-x}$Te$_4$ with Calcium Metal Substitution

Samantha M. Clarke[1,2]*, James M. Ma[1,3]*, C.-K. Huang[1], Paul A. von Allmen[1], Trinh Vo[1], Richard B. Kaner[2,3], Sabah K. Bux[1], and Jean-Pierre Fleurial[1]**

[1]Jet Propulsion Laboratory, California Institute of Technology, Pasadena, CA 91109, U.S.A.
[2]Department of Chemistry and Biochemistry, University of California, Los Angeles, CA 90092, U.S.A.
[3]Department of Materials Science and Engineering, University of California, Los Angeles, CA 90092, U.S.A.
*These authors contributed equally to this paper

## ABSTRACT

La$_{3-x}$Te$_4$ is a state-of-the-art high temperature n-type thermoelectric material with a previously reported maximum zT~1.1 at 1273 K. Computational modeling suggests the La atoms play a crucial role in defining the density of states for La$_{3-x}$Te$_4$ in the conduction band. In addition to controlling charge carrier concentration, substitution with Ca$^{2+}$ atoms on the La$^{3+}$ site is explored as a potential means to tune the density of states and result in larger Seebeck coefficients. High purity, oxide-free samples are produced by ball milling of the elements and consolidated by spark plasma sintering. Powder XRD and electron microprobe analysis are used to characterize the material. High temperature thermoelectric properties are reported and compared with La$_{3-x}$Te$_4$ compositions. A maximum zT of 1.3 is reached at 1273 K for the composition La$_{2.22}$Ca$_{0.775}$Te$_4$.

## INTRODUCTION

Thermoelectric (TE) power sources have consistently demonstrated their extraordinary reliability and longevity in support of the National Aeronautics Space Administration's (NASA) deep space science and exploration missions. Proven state-of–practice "heritage" (TE) materials exhibit only modest thermal-to-electric energy conversion performance, resulting in relatively low system-level conversion efficiencies of 6 to 6.5%. These heritage materials have been known since the late 1950's and 1960's, so that even the recently developed multi-mission radioisotope thermoelectric generator (MMRTG) builds upon 40-year old thermoelectric converter technology. However, there is great potential for large gains in performance thanks to recent advances in materials synthesis, the discovery of novel complex structure compounds, the ability to engineer with increasing precision micro- and nanostructure features coupled with improved scientific understanding of electrical and thermal transport in such engineered materials and the means to perform in-depth theoretical simulations with fast turnaround time.

NASA's Radioisotope Power Systems Technology Advancement Program is pursuing the development of more efficient thermoelectric technologies that can increase performance by a factor of 2 to 4X over state-of-practice systems. The use of advanced materials, n-type and p-type filled skutterudites and rare earth compounds, has already resulted in doubling TE couple-level conversion efficiency up to 15% at the beginning of their life. One of these rare earth compounds that demonstrates a high performance at high temperatures is lanthanum telluride, La$_{3-x}$Te$_4$, self-doped through La vacancies, with a reported zT of 1.1 at 1273 K when x = 0.23.(*1*)

ZT, defined as $zT = \frac{\alpha^2}{\kappa\rho}T$, is the dimensionless thermoelectric figure of merit, and directly impacts conversion efficiency. Good TE materials need to exhibit low thermal conductivity ($\kappa$), low electrical resistivity ($\rho$), and large magnitude of the Seebeck coefficient ($\alpha$). The thermal conductivity is the sum of its electronic contribution, $\kappa_e$, and its lattice contribution, $\kappa_L$.

$La_{3-x}Te_4$ possesses glass-like $\kappa_L$ values, and reasonably good electrical transport properties, with Seebeck coefficient and electrical resistivity values near 200 $\mu V.K^{-1}$ and 30 $\mu\Omega.m$, respectively, at 1273 K. Electron concentration is controlled by adjusting the number of La vacancies, with allowed x values ranging from x = 0 (metallic behavior) to x=0.33 (intrinsic semiconductor behavior).

From computational modeling and previous experimentation, it has been found that the density of the states (DOS) in the conduction band is dominated by the lanthanum atoms.(2) Non-isoelectronic substitutions, such as $Yb^{2+}$ on the $La^{3+}$ site, could potentially allow for the modification of the DOS, which could increase the Seebeck coefficient for this material.(3) The use of a divalent cation would allow for a finer control over the carrier concentration. The electronic local environment becomes $La_{3-x-y}^{3+} \blacksquare_{La,x} M_y^{2+} Te_4^{2-} e_{1-3x}^{1-}$, where $\blacksquare_{La,x}$ represents the number of vacancies in the system and $M_y^{2+}$ represents the amount of non-isoelectronic metal substituents in the system. The theoretical carrier density can be calculated by $n = n_{max}(1 - 3x - y)$ where $n_{max}$ =4.5×10²¹ cm⁻³.(3) There is a threefold finer control of carrier concentration in the $M^{2+}$ doped system compared to the vacancy-doped system. This makes it easier to obtain samples with carrier concentrations in the optimized region of $0.8x10^{21}$ cm⁻³ to $1.2x10^{21}$ cm⁻³, which had previously been difficult to obtain.(1) This doping also provides a way to create vacancy-free structures with identical carrier concentrations to the vacancy-doped samples, allowing for a separation of the impact on the lattice thermal conductivity of various defects on the La sublattice (vacancies, divalent atoms) versus electron-phonon interactions.

For this study, calcium was chosen as a dopant because of its definitive 2+ charge, its similarity in size to $La^{3+}$ which enhances the probability of successful substitution upon that site, its potential for lattice thermal conductivity reduction through point defect phonon scattering (much lower atomic mass), and its difference in electronic properties from La atoms as it lacks electrons in the d and f orbitals.

## EXPERIMENT

Samples were prepared using powder metallurgy from the elements in an argon glove box as previously described.(4) The powders were then compacted using graphite dies (POCO) and spark plasma sintering (SPS) under vacuum at temperatures above 1250 °C and a pressure of 80 MPa yielding samples with > 97% of theoretical density.

A Bruker D8 powder X-ray diffractometer using Cu Kα radiation was used to confirm phase purity and calculate the lattice parameter. A JEOL Super Probe electron microprobe analyzer and wave dispersive spectroscopy (WDS) was used to examine elemental compositions. The thermoelectric transport properties of the samples were measured using both custom setups and commercial instrumentation described elsewhere.(5, 6)

## DISCUSSION

### Characterization

A comparison of the X-ray diffraction patterns of the baseline $La_{2.8}Te_4$ sample to various compositions of $La_{3-x}Ca_yTe_4$ indicated that no second phases could be detected (figure 1). There is little change in the lattice parameter for the various Ca containing alloys, which is expected since the ionic radii of $Ca^{2+}$ (114.0 pm) is similar in size to the ionic radii of $La^{3+}$ (117.2 pm).

The Ca-substituted samples were found to be highly homogeneous and single phase as seen through the electron microprobe backscattered electron (BSE) image of a typical sample (figure 2). The wavelength dispersive spectroscopy (WDS) provided a way to determine the experimental stoichiometric amounts of each element and the corresponding electron count and carrier concentration (table 1).

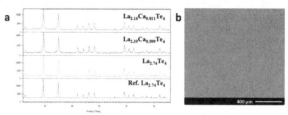

**Figure 1. a)** Powder X-ray diffraction comparing $La_{2.15}Ca_{0.811}Te_4$ (top row), $La_{2.39}Ca_{0.599}Te_4$ (second row), the vacancy doped $La_{2.8}Te_4$ (third row), and a reference pattern of lanthanum telluride (fourth row). All samples are single phase and no significant change in lattice parameter is observed. **b)** Image of $La_{2.39}Ca_{0.599}Te_4$ taken in backscattered electron mode showing homogeneous a sample. Slight color variations are attributed to porosity and/or surface oxidation.

**Table 1.** Summary of room temperature electrical transport properties. The calcium-doped system follows the similar trends as the vacancy doped system at room temperature. In the following plots, the samples will be referenced by their respective Hall carrier concentrations.

| Stoichiometry | Hall Carrier Conc. (cm$^{-3}$) | WDS Carrier Conc. (cm$^{-3}$) | Carrier Mobility (cm$^2$/Vs) | Electrical Resistivity (m$\Omega$*cm) | Seebeck ($\mu$V/K) |
|---|---|---|---|---|---|
| $La_{2.39}Ca_{0.599}Te_4$ | $2.5 \times 10^{21}$ | $1.6 \times 10^{21}$ | 2.3 | 1.09 | -32.3 |
| $La_{2.39}Ca_{0.638}Te_4$ | $1.4 \times 10^{21}$ | $2.0 \times 10^{21}$ | 4.6 | 0.96 | -35.1 |
| $La_{2.25}Ca_{0.778}Te_4$ | $1.3 \times 10^{21}$ | $1.4 \times 10^{21}$ | 3.8 | 1.28 | -37.2 |
| $La_{2.22}Ca_{0.775}Te_4$ | $6.7 \times 10^{20}$ | $1.0 \times 10^{21}$ | 3.3 | 2.85 | -61.3 |
| $La_{2.16}Ca_{0.811}Te_4$ | $5.5 \times 10^{20}$ | $5.2 \times 10^{20}$ | 4.8 | 2.37 | -50.7 |

### Transport Properties

The room temperature properties are summarized in table 1. The compositions are determined using the measured Hall carrier concentration and the calculated carrier concentration from WDS. A good agreement between the two is observed. The room temperature electrical resistivity and Seebeck coefficient of the Ca-substituted samples increases with decreasing carrier concentration leading to more semiconducting behavior which is consistent with $La_{3-x}Te_4$ compositions when the number of vacancies is increased.

The temperature dependence of the Seebeck coefficient and electrical resistivity for each of the samples is shown in figure 3a and 3b, respectively. The electronic property measurements indicate degenerate heavily doped semiconductor behavior with increasing resistivity and Seebeck with temperature. The calcium doped samples follow similar trends to $La_{3-x}Te_4$ at higher carrier concentrations. This is highlighted by the overlapping resistivity and Seebeck curves for $La_{2.8}Te_4$ (line) and $La_{2.39}Ca_{0.638}Te_4$ (triangles) at the similar carrier concentrations of $1.6x10^{21}$ $cm^{-3}$ and $1.4x10^{21}$ $cm^{-3}$, respectively.

There may be some indication that a modification to the density of states is occurring in the Ca-doped samples at lower carrier concentrations. This is observed in figure 3a and 3b in which the $La_{2.22}Ca_{0.775}Te_4$ ($6.7x10^{20}$ $cm^{-3}$) sample reaches a Seebeck coefficient of -300 $\mu V.K^{-1}$ with a resistivity of ~11 $m\Omega$.cm (110 $\mu\Omega$.m) at 1250 K. Samples prepared by May *et al*. show similar Seebeck values of -300 $\mu V/K$, however with a significantly higher resistivity ~26 $m\Omega$.cm (260 $\mu\Omega$.m) at 1250 K.([1]) This result may indicate that there may be some enhancement of the Seebeck coefficient at low carrier concentrations. However, more samples in this carrier concentration range coupled with first-principles electronic band structure calculations are needed to confirm this result.

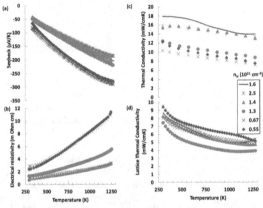

**Figure 3.** The La vacancy-doped $La_{2.8}Te_4$ is indicated by the solid line and the calcium-doped samples are indicated by the symbols. The samples are distinguished by their carrier concentrations as shown in Table 1. **a)** The Seebeck coefficient, where the solid line is completely covered by sample 1.4 represented by triangles. **b)** Electrical resistivity, the solid line is covered by triangles. **c)** Total thermal conductivity and **d)** lattice thermal conductivity.

The thermal conductivity (figure 3c) and lattice thermal conductivity (figure 3d) as a function of temperature are also shown. The heat capacity is obtained by altering the $La_{2.8}Te_4$ values measured by DSC by the calculated Dulong Petit value of each sample. Thermal conductivity was calculated using measured thermal diffusivity, heat capacity and thermal expansion from previously published DSC results, and were adjusted for calcium content using the Dulong Petit law.([7]) The electronic contribution was calculated in the first place using the Wiedemann-Franz law and adjusting the Lorenz number (2.2 is used) for degeneracy (using the Fermi level extracted from the Seebeck coefficient values). The minimum lattice thermal conductivity is achieved at a carrier concentration of around $1 \times 10^{21}$ cm$^{-3}$ for the calcium-doped system. At 300 K, the thermal conductivity for La$_{2.22}$Ca$_{0.775}$Te$_4$ at a carrier concentration of $6.7 \times 10^{20}$ cm$^{-3}$ is 15.4 mW.cm$^{-1}$.K$^{-1}$ (1.54 W.m$^{-1}$K$^{-1}$) which is approximately 14% lower than the La vacancy-doped La$_{2.8}$Te$_4$ at a similar carrier concentration. This reduction in the lattice thermal conductivity is likely due to point defect scattering by the lighter Ca$^{2+}$ ions in the La$_{3-x}$Te$_4$ structure. The effect of point defect scattering is diminished at higher temperatures (>850K) and the thermal conductivities are comparable. This mechanism could potentially be utilized to improve zT across the lower temperature range and enhance the thermoelectric device efficiency. A more comprehensive set of Ca-substituted samples, with and without La vacancies need to be studied to be able to precisely ascertain the relative impact of La vacancies, Ca substitutions, and carrier concentration on decreasing lattice thermal conductivity values.

With the Ca substitutions, samples with carrier concentrations in the optimized region were fairly easily synthesized, which had been difficult to achieve previously.(1) May et al. reports obtaining carrier concentrations of $0.45 \times 10^{21}$ cm$^{-3}$ and $1.6 \times 10^{21}$ cm$^{-3}$, however the authors commented that samples with carrier concentrations lower than $1.6 \times 10^{21}$ cm$^{-3}$ are difficult to synthesize and test since they didn't readily sinter and were subject to rapid oxidation .(1) In the calcium-doped samples reported here, carrier concentrations of $1.3 \times 10^{21}$ cm$^{-3}$ and $0.67 \times 10^{21}$ cm$^{-3}$ were obtained and are much closer to the predicted optimized carrier concentration of $0.9 \times 10^{21}$ cm$^{-3}$. A potential explanation is that the Ca$^{2+}$ substitutions add stability to the crystal structure over the vacancies which make synthesis of samples in this range more feasible.

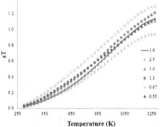

**Figure 5.** The TE figure of merit (zT) as a function of temperature for all samples. A peak zT of 1.3 at 1250 K is observed for La$_{2.22}$Ca$_{0.775}$Te$_4$ at a carrier concentration of $6.7 \times 10^{20}$ cm$^{-3}$.

Through Ca substitution, a finer control over carrier concentration is realized. The La$^{3+}$:Ca$^{2+}$ composition can be adjusted to only add one electron at a time to the system, whereas the vacancy doped system adds three. This is realized in the ability to synthesize samples in the $\sim 1 \times 10^{21}$ cm$^{-3}$ range. Future experiments will build upon this capability and additional samples in this carrier concentration range will be synthesized. Fine carrier concentration control is important for the La$_{3-x}$Te$_4$ system as near the optimized carrier concentration small deviations result in significant changes in the thermoelectric performance.

The zT as a function of temperature is shown in figure 5. Similar carrier concentrations, La$_{2.8}$Te$_4$ and La$_{2.39}$Ca$_{0.638}$Te$_4$ have zT values which are equivalent at all temperatures and reach a maximum of zT~1.1 at 1273 K. For calcium doped lanthanum telluride at lower carrier concentrations, La$_{2.22}$Ca$_{0.775}$Te$_4$ with a carrier concentration of $6.7 \times 10^{20}$ cm$^{-3}$ there was an overall increase in zT of 1.3 at 1273 K, or about 15% across the entire temperature range of interest. The calcium-doped system reaches higher maximum zT values than the La vacancy-doped system for

87

lower carrier concentrations when compared to the data presented by May et al.(*1*) This is predicted to be due to the increased Seebeck coefficient and decreased thermal conductivity at equivalent carrier concentrations. Preliminary evidence suggests that the calcium substitutions have potentially altered the density of states, but further experimental work, coupled with first – principles electronic band structure calculations, is necessary in the optimized region of the carrier concentration to validate the result.

## CONCLUSIONS

The successful doping of $La_{3-x}Te_4$ with $Ca^{2+}$ was achieved through powder metallurgy and indicates that substitutions of $Ca^{2+}$ for $La^{3+}$ is possible for a wide range of carrier concentrations. General behavior of the material is similar to that of La-vacancy controlled doping of $La_{3-x}Te_4$ first described by May et al. Preliminary evidence suggests an improvement in zT through the following mechanisms.(*1*) First, there is a potential enhancement of the Seebeck coefficient through modification of the density of states, observed at lower carrier concentrations.(*2*) Second, finer control over carrier concentration is achieved by having a divalent cation substituted system rather than a La vacancy-doped system. Additionally, the definitive oxidation state of $Ca^{2+}$ eliminates any ambiguity introduced from doping with mixed valent $Yb^{2+}/Yb^{3+}$.(*3*) Fourth, the lattice thermal conductivity is reduced at low temperatures via point defect scattering.

Further experimental work with Ca-and Ba- substituted samples, coupled with first – principles electronic band structure calculations, are planned to deconvolute the point defect and electronic transport property mechanisms at play in this promising thermoelectric materials system.

## ACKNOWLEDGMENTS

This work was performed at the Jet Propulsion Laboratory, California Institute of Technology under contract with the National Aeronautics and Space Administration. This work was supported by the NASA Science Missions Directorate's Radioisotope Power Systems Technology Advancement Program. Support was also provided in part by NSF IGERT: Materials Creation Training Program (MCTP) – DGE-0654431, the California NanoSystems Institute, the summer Undergraduate Research Fellowship (SURF) through CalTech and JPL, and the UCLA MSD Scholars Program. The authors would also like to thank Frank T. Kyte for his indispensable assistance and training in electron microprobe analysis.

## REFERENCES
1. A. May, J.-P. Fleurial, G. Snyder, *Physical Review B* **78**, 1–12 (2008).
2. A. May, D. Singh, G. Snyder, *Physical Review B* **79**, 1–4 (2009).
3. A. F. May, J.-P. Fleurial, G. J. Snyder, *Chemistry of Materials* **22**, 2995–2999 (2010).
4. A. May, J. Snyder, J.-P. Fleurial, M. S. El-Genk, *AIP Conference Proceedings* **969**, 672–678 (2008).
5. J. A. Mccormack, J. Fleurial, *Materials Research Society Symposium Proceedings* **234**, 135–143 (1991).
6. C. Wood, D. Zoltan, G. Stapfer, **56**, 719–722 (1985).
7. O. Delaire *et al.*, *Physical Review b* **80**, 1–9 (2009).

Mater. Res. Soc. Symp. Proc. Vol. 1490 © 2012 Materials Research Society
DOI: 10.1557/opl.2012.1670

# Anisotropy and inhomogeneity measurement of the transport properties of spark plasma sintered thermoelectric materials

A. Jacquot[1], M. Rull[2], A. Moure[3], J.F. Fernandez-Lozano[3], M. Martin-Gonzalez[2], M. Saleemi[4], M.S. Toprak[4], M. Muhammed[4], M. Jaegle[1]

[1]Fraunhofer-IPM, Thermoelectric Systems department, Heidenhofstraße 8, 79110 Freiburg, Germany.
[2]Instituto de Microelectrónica de Madrid, C/ Isaac Newton 8. Tres Cantos, 28760 Madrid, Spain.
[3]Instituto de Ceramica y Vidrio, C/ Kelsen, 5 Madrid 28049, Spain.
[4]Functi Instituto de Ceramica y Vidrio onal Materials Division, KTH Royal Institute of Technology, Kista-Stockholm, Sweden.

## ABSTRACT

We report on the development and capabilities of two new measurement systems developed at Fraunhofer-IPM. The first measurement system is based on an extension of the Van der Pauw method and is suitable for cube-shaped samples. A mapping of the electrical conductivity tensor of a Skutterudite-SPS samples produced at the Instituto de Microelectrónica de Madrid is presented. The second measurement system is a ZTmeter also developed at the Fraunhofer-IPM. It enables the simultaneous measurement of the electrical conductivity, Seebeck coefficient and thermal conductivity up to 900 K of cubes at least $5 \times 5 \times 5$ mm$^3$ in size. The capacity of this measurement system for measuring the anisotropy of the transport properties of a $(Bi,Sb)_2Te_3$ SPS sample produced by KTH is demonstrated by simply rotating the samples.

## INTRODUCTION

Thermoelectric materials enter in the fabrication of thermoelectric generators and coolers. The conversion efficiency is a monotonic growing function of the figure of merit defined as $Z = \sigma S^2 / \lambda$, where $\alpha$ is the Seebeck coefficient, $\sigma$ the electrical conductivity and $\lambda$ the thermal conductivity. These transport properties are taken all in the same direction. Larger electrical conductivity and Seebeck coefficient in conjunction with a low thermal conductivity results in useful thermoelectric materials. Since some of the best thermoelectric materials do have an anisotropic crystal structure and since plasma sintering (SPS) is uniaxial most or all samples produced by SPS should show to some extend an anisotropy of the transport properties. In addition, inhomogeneity is also expected in SPS samples because the current density, the temperature field and pressure may not be homogeneous during the sintering process. For the above mentioned reasons, it would be useful to map the transport properties of SPS-samples measured along and perpendicular to the sintering direction. Nevertheless, the thermal conductivity is more conveniently measured along a direction which is perpendicular to the measurement direction of the electrical properties with the measurement systems actually available on the market. This methodology may lead to an overestimation of the figure of merit in the case of a strongly anisotropic crystal structure [1-2]. These overestimated figure-of-merits will never be translated into any improvement of the conversion efficiency in real systems. In addition, lots information about physical phenomena could be grasped if the anisotropy of the transport properties were more easily measured [3-6]. Nevertheless this is not an easy task because the optimal geometry for the electrical properties and thermal conductivity measurements are usually different. It has also been noticed that an overestimation of the figure of merit may arise when the samples are inhomogeneous or thermally not stable [7]. Therefore it would be very desirable to measure all the transport properties on the same sample in the same direction.

In this article, we are presenting two solutions for the measurement of the anisotropy of transport properties. The first solution is an extension of the of the Van der Pauw method [8] for cubic samples (3D-VdP). The method presented differs markedly in its simplicity with previously published papers [9-15]. This solution is nevertheless restricted to the measurement of the electrical conductivity. The second solution is not restricted to the electrical conductivity and is based on the IPM-ZTmeter [16].

## EXPERIMENTAL DETAILS

### 3D-Van der Pauw method

The starting material used to test the method was a Skutterudite based wafer provide by Instituto de Microelectrónica de Madrid in spain (wafer ID . CSIC-SPS-641). It was not expected to get any anisotropy of the electrical conductivity because of the crystal structure but the sintering process which is uniaxial can introduce some anisotropy as well as inhomogeneity in the transport properties. The wafer was cut into cubes of size 2x2x2mm3 has shown in the insert of the figure 4. Usual VdP measurements have been performed on the face X and Y of each carefully oriented cube cubes (figure 1). Resistance measured on the face of the cube Y along the X axis was labeled $R_{X,Face\,Y}$. Assuming an isotropic electrical conductivity perpendicular to the sintering direction (parallel to the Face Y), the resistance $R_{X,Face\,Y}$ and $R_{Z,Face\,Y}$ must be equal. The result of the measurement in this case is a unique resistance value $\left(R_{X\|Z,Face\,Y}\right)$.

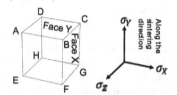

**Figure 1.** Conventions and labeling used in conjunction with the 3D-VdP method.

When the VdP measurement is performed on the Face X (or Z), two distinct resistance values are obtained $R_{X\|Z,Face\,Y} \neq R_{X\|Z,Face\,X}$. In fact it is sufficient to perform the VdP measurement on the face X (or Z) in order to extract the electrical conductivity along the sintering direction $(\sigma_Y)$ and perpendicular to it $(\sigma_{XZ})$. This is because the resistance is a function of $\sigma_Y$, $\sigma_{XZ}$ and the geometry (edge length), the latest being known. $\sigma_Y$ and $\sigma_{XZ}$ are obtained by reporting the measured resistance values in the R-$\sigma$ graphs (figure 2) calculated using Comsol Multiphysics [17-18] for the particular geometry and measurement configuration. If a measurement is additionally made on the face Y, it is only used to confirm the results obtained with the face X or Z. The error in the electrical conductivity measurement is less than 5%, far less than with the ZTmeter. The key advantage of the VdP method is that punctual contacts are used. The placement of the measurement probes and geometric imperfections introduced when cutting the samples are the major sources of error in the 3D-VdP method. Error bar may increase with larger cubes and larger electrical conductivity.

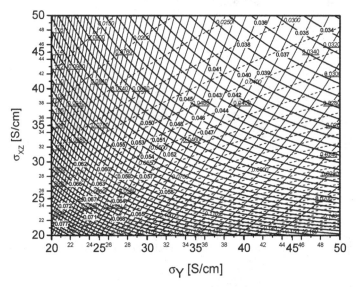

**Figure 2.** R-$\sigma$ graphs used to extract the electrical conductivity of the sample CSIC-SPS-641 along and perpendicular to the sintering direction. It is the superposition of three R-$\sigma$ diagrams. The labels inside the plot area are resistances calculated for specific faces and direction of a 2x2x2mm³ sample. The bold solid lines and bold labels represent the contour lines and value of $R_{X\|Z,FaceY}$, respectively. The thinner solid lines and underlined labels correspond to the contour lines and value of $R_{Y,FaceX\|Z}$. Finally, the dashed lines and label in italic stand for $R_{X\|Z,FaceX\|Z}$.

### IPM-ZTMeter

All the transport properties were measured in the steady state regime, on the same sample and with a temperature difference along the sample of few Kelvins. The heat flows were measured above and below the sample in order to be able to subtract the heat loss by radiation. In its current state of development the error bars for the ZTmeter are large especially for the electrical and thermal conductivity. Variability in the measurements is often the result of inhomogeneous electrical and thermal contacts. It should be noted nevertheless that the propagation of uncertainty is rather favorable to the ZTmeter concerning the Seebeck coefficient. The error in the Seebeck coefficient propagates in the ZT rather linearly than with a quadratic form because the same temperature difference along the sample is used for the Seebeck and thermal conductivity measurement.

91

The sample measured (KTH_SPS#16352), 5x5x5 mm$^3$ in size, was nanopowder Bi$_{1.9}$Sb$_{0.1}$Te$_3$ synthetized and sintered at the Royal Institute of Technology in Sweden (KTH). The morphology of the powder is shown in the figure 3. Details about the powder production and compaction can be found in [19].

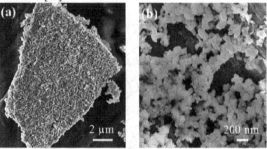

**Figure 3.** Morphology of the powder that enter in the fabrication of the sample KTH-SPS-16252

## DISCUSSION

### Mapping of the electrical conductivity and its anisotropy

Obviously the electrical conductivity of the wafer is inhomogeneous (figure 4). It seems also that the electrically is slightly larger along the sintering direction by few percent. Nevertheless the anisotropy measured is very small and below the measurement uncertainties.

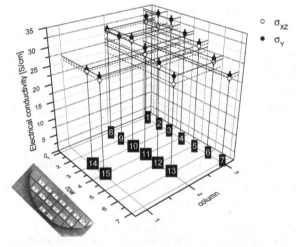

**Figure 4.** Mapping of the electrical conductivity measured along and perpendicular to the sintering direction of the sample CSIC-SPS-641. The labels with the black background correspond to the labels of the cubes shown in the insert.

## Measurement of the anisotropy of the transport properties at high temperatures

The sample KTH_SPS#16352 is n-type (figure 5) as expected for this particular chemical composition [20]. N-type soft super lattice based on $Bi_2Te_3/(Bi,Sb)_2Te_3$ created by nanoalloying as also been observed [21]. The anisotropy of the material is very strong even if it was produced from nanopowder. This result is confirmed by a 3D-VdP measurement ($\sigma_{XZ}$ =800S/cm; $\sigma_Y$ =480 S/cm).The anisotropy of the Seebeck coefficient seems to indicate that there is both electrons and holes in the material, which also tend to explain the rather low ZT measured.

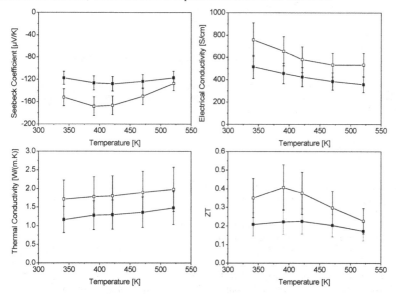

Figure 5: Transport properties measurement of the sample KTH-SPS-16252

## CONCLUSIONS

The 3D-VdP method enables the mapping of the electrical conductivity and its anisotropy The R-$\sigma$ graphs obtained for particular sample geometry frees the end users of having to perform themselves numerical simulations making the method easy to implement. Since it has been proved helpful to us in the purpose of investigating the homogeneity of SPS-samples and to validate the electrical conductivity measurement made with our ZTmeter, this method should find a widespread use. The IPM-ZTmeter has been proved able to measure on a single sample, the anisotropy the transport properties at high temperature using just 5x5x5 mm$^3$ materials. It has been demonstrated that the compaction of nanopowder does not lead systematically to isotropic materials.

## ACKNOWLEDGMENTS

The research work is supported by the European Commission under FP7-NEXTEC project, Grant # 263167 [FP7/2011-2013]. KTH_SPS#16352 was produced by Aleksey Ruditskiy who is now working at the City College of New York

## REFERENCES

1. A. Jacquot, J. Thomas, J. Schumann, M. Jägle, H. Böttner, T. Gemming, J. Schmidt and D. Ebling, J. Mater. Res. 26, 1773 (2011).
2. X. Yan, B. Poudel, Y. Ma, W. S. Liu, G. Joshi, H. Wang, Y. Lan, D. Wang, G. Chen and Z. F. Ren, Nano Lett. 10, 3373 (2010).
3. M. Situmorang and H.G. Goldsmid, Phys. Stat. Sol. (b) 134, K83 (1986).
4. A. Jacquot, B. Bayer, M. Winkler, H. Böttner, M. Jaegle, J. Solid State Chem. 193, 105 (2012).
5. A. Jacquot, J. König, B. Bayer, D. Ebling, J. Schmidt, M. Jaegle, Coupled theoretical and experimental investigation of the role of impurity levels and concentration in Bi$_2$Te$_3$ and PbTe based materials at high temperature, 8$^{th}$ European Conference on Thermoelectrics, Como, September 22-24, P1-12 (2010).
6. M. Storder, H.T. Langhammer, phys. Stat. sol (b) 117, 329 (1983).
7. Von H. Fleischmann, Z. Naturforschg 16 a, 765 (1961).
8. L.J. van der Pauw, Philips Res. Repts 13, 1 (1958).
9. D.W. Koon, Rev. Sci. Instrum. 60, 271 (1989).
10. C. Kasl and M.J.R. Hoch, Rev. Sci. Instrum. 76, 033907 (2005).
11. I. Kazani, G. De Mey, C. Hertleer, J. Banaszczyk, A. Schwarz, G. Guxho and L. Van Langenhove, Text. Res. J. 0(01) 1 (2011).
12. J. Kleiza and V. Kleiza, Lith. J. Phys. 45 N°5 333 (2005).
13. Y. Sato and S. Sato, Jpn. J. Appl. Phys. Vol. 40, 4256 (2001).
14. O. Bierwagen, R. Pomraenke, S. Eilers, and W. T. Masselink, Phys. Rev. B 70, 165307 (2004).
15. J. de Boor, V. Schmidt, Adv. Mater. 22, 4303 (2010).
16. A. Jacquot, M. Jaegle, H.-F. Pernau, J. König, K.Tarantik, K. Bartholomé, H. Böttner, "Simultaneous measurement of the thermoelectric properties with the new IPM-ZTMeter", 9th European Conference on Thermoelectrics, Thessaloniki, September 28-30, C-14-P (2011).
17. M. Jaegle, M. Bartel, D. Ebling, A. Jacquot and H. Böttner, "Multiphysics simulation of thermoelectric systems", Proceedings of the 6$^{th}$ European Conference on Thermoelectrics, July 2-4, Paris, 0-27-1 (2008).
18. D. Ebling, M. Jaegle, M. Bartel, A. Jacquot and H. Böttner, Journal of Electronic Materials 38 Issue 7, 1456 (2009).
19. M. Saleemi, M. S. Toprak, S. Li, M. Johnsson and M. Muhammed, J. Mater. Chem. 22, 725 (2012).

20. R. Martin-Lopez, B. Lenoir, A. Dauscher, H. Scherrer and S. Scherrer, Solid State Commun. 108 N°5, 285 (1998).
21. M. Winkler, D. Ebling, H. Böttner, L. Kirste, "Sputtered n-type Soft Super Lattice Thermoelectric Layers Based on $Bi_2Te_3$/$(Bi,Sb)_2Te_3$ Created by Nanoalloying", Proceedings of the 8th European Conference on Thermoelectrics, September 22-24, Como, 26 (2010).

Mater. Res. Soc. Symp. Proc. Vol. 1490 © 2012 Materials Research Society
DOI: 10.1557/opl.2012.1557

# Formation and Properties of TiSi$_2$ as Contact Material
# for High-Temperature Thermoelectric Generators

Fabian Assion, Marcel Schönhoff and Ulrich Hilleringmann
Department of Sensor Technology, University of Paderborn, 33098 Paderborn, Germany

## ABSTRACT

Thermoelectric generators (TEG) are capable of transforming waste heat directly into electric power. With higher temperatures the yield of the devices rises which makes high-temperature contact materials important. The formation of titanium disilicide (TiSi$_2$) and its properties were analyzed and optimized for the use in TEG. Depending on a direct or an indirect transformation into the C54 crystal structure the process forms a layer with a resistivity of 20-22 µΩcm. Process gases influence the resistivity and result in difference of 20%. The growing rate of TiSi$_2$ on silicon dioxide was determined; it shows a strong dependence on the used atmosphere and temperature. A maximum overgrowing length of 30 µm was found.

## INTRODUCTION

It is estimated that between 20 to 50% of industrial energy input is lost as waste heat in the form of hot exhaust gases, cooling water, and heat lost from hot equipment surfaces and heated products [1]. A thermoelectric generator (TEG) can transfer this heat directly into electrical energy without any moving parts. Therefore, those generators do not need any kind of maintenance or extra fuel which makes every single watt output power to a profit without subtractions. For a fast gain back of the investment costs and a better environmental balance the TEG efficiency should be as high as possible. The efficiency of TEG rises with the applied temperature difference across its thermocouples. Hence, it is reasonable to increase the thermal stability of such devices and to expand the field of applications to higher temperatures. Aiming on thermally stable contacts we investigate in titanium disilicide (TiSi$_2$), which has a resistivity of 15-35 µΩcm [2,3] and withstands temperatures up to 1150 K [4]. Thereby TiSi$_2$ fulfills the requirements for an implementation of TEGs at the exhaust manifold of usual street cars where a temperature difference versus cooling water of up to 700 K could be applied (Figure 1).

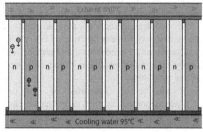

Figure 1: Schematic of a thermoelectric generator for an automotive application

## EXPERIMENT

The experiments covered in this paper deal with the influence on the $TiSi_2$ sheet resistance depending on different process parameters and the $TiSi_2$ overgrowth on silicon dioxide films $(SiO_2)$. A 200 nm titanium layer was deposited by magnetron sputtering on silicon wafers. After structuring the titanium layer it was heated up by rapid-thermal-annealing (RTA). This step forms titanium disilicide at between 600 °C till 700 °C in the instable C49 crystal structure (with resistivity of 60-300 $\mu\Omega$cm [2]). Higher temperatures let the base-centered-orthorhombic C49 crystal structure directly switch to the thermodynamically favored C54, which has a face-centered-orthorhombic structure [3]. Table I is giving more details about the crystal structure. To be able to insert $TiSi_2$ into FEM-simulations and design efficient TEG with it Table II sums up the several material properties of the silicide.

Table I: Lattice constants in angstrom [5]

| Compound | a | b | c |
|---|---|---|---|
| TiSi$_2$ (C49) | 13,76 | 3,61 | 3,85 |
| TiSi$_2$ (C54) | 8,55 | 4,79 | 4,07 |

Table II: Material properties of Titanium disilicide

| Propertie | Value | Reference | Propertie | Value | Reference |
|---|---|---|---|---|---|
| Density | 4,07 g/cm³ | [6] | Melting point | 1773 +/-10 K | [7,9,11] |
| Young's modulus | 200-355 GPa | [7,8,9] | Thermal stability | 1150 K (in air) | [4] |
| Poisson's ratio | 0,22-0,24 | [8,10] | Thermal conductivity | 40,5-29,3 W/mK @573-1773 K | [9] |
| Compressive strength | 117,9 MPa @293 K 5,5 MPa @1473 K | [7] | Thermal expansion | 12,5 · 10$^{-6}$ 1/K 6,403 + 0,007267T · 10$^{-6}$ 1/K | [7] [12] |
| Tensile strength | 150 MPa @293 K | [7,9] | Spezific heat | 51,47 J/molK @298,15 K 102,96 J/molK @1200 K | [9] |
| Bend strength | 210 MPa @293 K | [9] | Schottky Barrier | 0,60 eV | [13] |

The next step is the wet etching of the remaining, not reacted titanium and titanium nitride. The second RTA process at temperatures above 700 °C changes the crystal structure of the $TiSi_2$ into the C54 formation, which has a lower resistivity and a higher thermal stability. The transformation temperature has been shown to be a function of film thickness [3] and dopants in the silicon [14]. In different test series time, temperature and atmosphere of the RTA processes have been variegated. Additionally three different widths of the $TiSi_2$ structures were used to see if there is a dependence of the edge roughness.

The second test was done to analyze the overgrowth of $TiSi_2$ on silicon dioxide films. Therefore, a 200 nm $SiO_2$ layer was deposited by plasma enhanced chemical vapor deposition.

After etching windows into the oxide layer 200 nm of titanium were deposited like in the first test. Like the resistivity tests different test series were done to check the dependence of time, temperature and atmosphere of the RTA process.

## DISCUSSION

On-wafer electrical characterization was performed using an HP 4156A precision semiconductor parameter analyzer. Four point measurements have been used to minimize influences of the measurement system itself.

The diagrams 2a and b show the resistivity of the $TiSi_2$ during the RTA process. Whereas the first dots represent the resistivity after the first annealing step which is a mixture of $TiSi_2$ C49, not-reacted titanium and titanium nitride. The value after the etching step is given in the second row. The last value is the resistivity of $TiSi_2$ C54 after the second annealing. Average values for $TiSi_2$ C49 as well as the literature maximum and minimum value for $TiSi_2$ C54 are also marked in the diagram. Figure 2a presents the results for the RTA process in argon (Ar) atmosphere. The resistivity is falling during the process as indented. Values in each row are widely spread because the Ar atmosphere is not totally free of oxygen. Even though the chamber was evacuated below $5 \cdot 10^{-3}$ Pa before process gases were turned on the remaining oxygen reacts with the titanium and forms titanium oxide which is an insulator. The titanium oxide cannot be etched with ammonium hydroxide which is used for the removal of non-reacted titanium. Therefore, the titanium oxide remains and reduces the resistivity of the resulting layer.

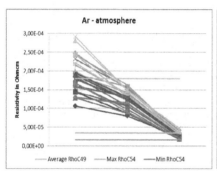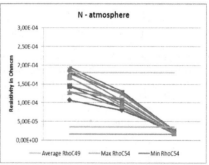

Figure 2: $TiSi_2$ resistivity during the process (after first annealing; etching and second annealing) in argon (a) and nitrogen (b) atmosphere

These measurements were done at the same temperatures of 650 °C for the first annealing and 750 °C for the second step. Longer process time enhances the titanium oxide formation, and therefore the sheet resistance increases. An enhancement from 60 s to 90 s results in a 30 % increased resistivity.

Figure 2b shows the results of the same process in nitrogen ($N_2$) atmosphere. It can clearly be seen that the spreading inside the single rows is much smaller. The reason for this is a reaction between the titanium and the nitrogen. It suppresses the oxidation during the process by forming titanium nitride inside the grain boundaries which reduces the oxygen diffusion. The titanium nitride can easily be removed and enhances the repeatability. In total the $N_2$ atmosphere reduces the resulting resistivity compared to an Ar atmosphere by 20% which results in 22 μΩcm. Opposite to the Ar process the variegation of the process time or the structure width showed no notable impact on the sheet resistance. These reasons make the nitrogen process superior.

The process temperature has a major influence on the out coming product in both atmospheres. This is because the temperature is the crucial factor for the crystal structure switching. Figure 3 shows results of a process with higher temperatures. When the first annealing step is done at 750 °C the TiSi$_2$ directly switches into the C54 crystal structure. So after etching the titanium nitride, the second annealing step shows nearly no more reaction. The average resistivity is even lower than in the processes described before. The extracted resistivity for TiSi$_2$ formed in a one-step-process is 20 μΩcm.

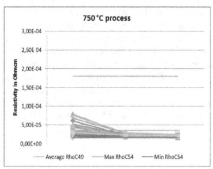

Figure 3: TiSi$_2$ resistivity during the process with 750 °C for the first annealing in N$_2$-atmosphere

The overgrowth of TiSi$_2$ over SiO$_2$ was observed via an alpha-step and a scanning electron microscope (SEM). Figure 4 shows the alpha-step-measurement and figure 5a the SEM-picture of an edge of a SiO$_2$ window with titanium on top of it before the second RTA process and after the etching. The step height of the diagram before the process equals the thickness of the SiO$_2$ layer. After etching the step height is lower because the not reacted titanium on top of the SiO$_2$ is removed. Nevertheless the overgrowth length is clearly recognizable.

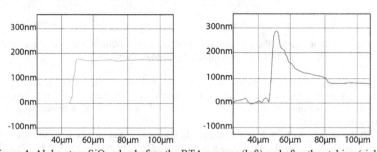

Figure 4: Alpha-step: SiO$_2$ edge before the RTA process (left) and after the etching (right)

When the temperature-time-budget is high enough $TiSi_2$ can be formed directly on $SiO_2$ which can be seen in figure 5b. IIDA and ABE showed that this transformation already starts at room temperature but needs hours to be detectable [15].

Figure 5: SEM-picture: $SiO_2$ edge after RTA (a: 90s@700°C / b: 420s@650°C) and etching

Figure 6a presents the results of two test series concerning the process dependence on the atmosphere. It can be seen that at a temperature of 650 °C the $TiSi_2$ growth over the $SiO_2$ is almost linear. The growing rate in nitrogen with 70 nm/s is twice as fast as in argon atmosphere with 35 nm/s. Concerning the atmosphere this result is different to former publications [2]. The growing rate contrariwise matches results from OKAMOTO [16]. He suggests a proportional growth to the square root of time, which means that there should be an asymptotic behavior. Figure 6b confirms this assumption and shows that the growing rate is strongly dependent on the process temperature.

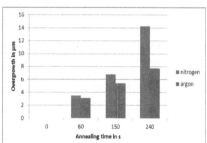 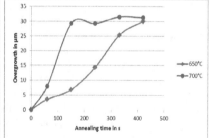

Figure 6: $SiO_2$ overgrowth depending on the (a) atmosphere and (b) temperature

Influences of the surface on which the $TiSi_2$ was formed have also been analyzed. Unexpectedly, the best results were found on rough instead of plane surfaces. Since silicon is the moving reaction agent, grooves do enhance the formation of $TiSi_2$ because more silicon is present. Consequently the grooves lead to a priority conductive direction. This fact was used in a $TiSi_2$ contacted demonstrator TEG which proofed a thermal stability up to 900 K and reduce the joined resistance of a former TEG by a factor of four [17]. By using amorphous silicon deposited via plasma enhanced chemical vapor deposition the technology is transferable onto almost any material. Analyzing the contact resistance between $TiSi_2$ and different other materials is part of current research efforts.

101

## CONCLUSIONS

The formation and the properties of Titanium disilicide have been analyzed and process dependencies were spotted and documented. The optimal two-step-process forms a $TiSi_2$ with an average resistivity of $22\,\mu\Omega cm$. The direct C54 process results in an average resistivity of $20\,\mu\Omega cm$. A resistivity difference of 20 % was found between the process gases argon and nitrogen. The growing rate of $TiSi_2$ on silicon dioxide in nitrogen with $70\,nm/s$ is twice as fast as in argon atmosphere ($35\,nm/s$). In both atmospheres a maximum overgrowing length of $30\,\mu m$ was found.

## ACKNOWLEDGMENTS

The author would like to thank the German Federal Ministry of Education and Research (BMBF) for founding the Project HOTGAMS (03X3547A).

## REFERENCES

1.   Energetics, Energy Use, Loss, and Opportunities Analysis: U.S M&M; pp. 17; 2004.

2.   E.G. Colgan et al; Materials Science and Engineering; 16; p. 43 – 96; 1996.

3.   Z. Ma, L. Allen and D. Allman; 'Microstructural aspects and mechanism of the C49-to-C54 polymorphic transformation in titanium disilicide'; J. Appl. Phys.; 77(8); 1995.

4.   S. P. Murarka; 'Self-aligned silicides or metals for very large scale integrated circuit applications' in Journal of Vacuum Science & Technology; 4 (6); p. 1325 – 1331; 1986.

5.   K. Maex and M. van Rossum; 'Properties of Metal Silicides'; INSPEC; 1995.

6.   NBS Monograph 25 (USA); no. 21; p. 126; 1984.

7.   G.V. Samsonov, I.M. Vinitskii; Handbook of Refractory Compounds; 1980.

8.   R. Rosenkrantz. G. Frommeyer; Z. Met. Kd. (Germany); vol. 83; no. 9; p. 685-689; 1992.

9.   T.V. Kosolapova; Handbook of High Temperature Compounds: Properties, Production, Application (Hemishere Publishing Corporation); 1990.

10.  P.J.J. Wessels, J.F. Jongste, G.C.A.M. Janssen, A.L. Mulder, S. Radelaar, O.B. Loopstra; J. Appl. Phys. (USA); vol. 63; no. 10; p. 4979-4982; 1988.

11.  T.B. Massalski; Binary Alloy Phase Diagrams, 2nd edition; vol.1-3; ASM; 1990.

12.  I. Engström, B. Lönnberg; J. Appl. Phys. (USA); vol. 63; p.4476-4484; 1988.

13.  J.Y. Duboz et al; Appl. Surf. Sci. (Netherlands) ; vol. 38; p. 171-177; 1989.

14.  E. Ganin, S. Wind, et. al; MRS Symposium Proceedings, vol. 303, Rapid Thermal and Integrated Processing 11, edited by Jeffrey C. Gelpey, J. Kiefer Elliott, et.al; MRS, Pittsburgh, PA; pp. 109; 1993.

15.  S. Iida; S. Abeb; Appl. Surface Sci.; Vol 78 (2), p 141–146, June 1994.

16.  T. Okamoto; K. Tsukamoto; M. Shimizuan and T. Matsukawa; 'Titanium silicidation by halogen lamp annealing'; J. Appl. Phys; 57 (12); pp. 5251-5255; 1985.

17.  F. Assion, M. Schönhoff and U. Hilleringmann: "Titaniumdisilicide as High-Temperature Contact Material for Thermoelectric Generators"; International Conferenc on Thermo- electrics, Aalborg, Denmark, 9-12 July, 2012.

Mater. Res. Soc. Symp. Proc. Vol. 1490 © 2013 Materials Research Society
DOI: 10.1557/opl.2013.117

# New n-type Silicide Thermoelectric Material with High Oxidation Resistance

Ryoji Funahashi[1, 4], Yoko Matsumura[1], Tomonari Takeuchi[1], Hideaki Tanaka[1], Wataru Norimatsu[2, 4], Emmanuel Combe[1], Ryosuke O. Suzuki[3, 4], Chunlei Wan[2, 4], Yifeng Wang[2, 4], Michiko Kusunoki[2, 4] and Kunihito Koumoto[2, 4]

[1]National Institute of Adv. Industrial Sci. and Tech., Midorigaoka, Ikeda, Osaka 563-8577, Japan
[2] Nagoya University, Nagoya, Aichi 464-8603, Japan
[3] Hokkaido University, Sapporo, Hokkaido 060-8628, Japan
[4] CREST, Japan Science and Technology Agency, Chiyoda, Tokyo 102-0075, Japan

## ABSTRACT

In order to achieve waste heat recovery using thermoelectric systems, thermoelectric materials showing high conversion efficiency over wide temperature range and high resistance against oxidation are indispensable. A silicide material with good n-type thermoelectric properties and oxidation resistance has been discovered. The composition and crystal structure of the silicide are found out $Mn_3Si_4Al_2$ (abbreviated as 342 phase) and hexagonal $CrSi_2$ structure, respectively. Element substitution of Mn with 3d transition metals is succeeded. Enhancement of Seebeck coefficient is observed in a Cr-substituted sample. The maximum dimensionless thermoelectric figure of merit $ZT$ is 0.3 at 573 K in air for the $Mn_{2.7}Cr_{0.3}Si_4Al_2$ sample. Electrical resistivity of the $Mn_3Si_4Al_2$ bulk sample holds constant value for 48 h at 873 K in air. This is due to formation of oxide passive layer on the surface of the bulk sample. The 342 phase is a promising n-type material with a good oxidation resistance in the middle temperature range of 500-800 K.

## INTRODUCTION

The demand for primary energy in the world was 12,013 million tons of oil per year in 2007 [1]. The average of total thermal efficiency of the thermal systems utilizing the fuel is limited to about 30%, with about 70% of the heat exhausted to the air as waste heat. It is clear that improving the efficiencies of these systems could have a significant impact on energy consumption. Electricity is a convenient form of energy that is easily transported, redirected, and stored; thus there are a number of advantages to converting the waste heat emitted from our living and industrial activities to electricity. Thermoelectric conversion is attracting attention because it is the strongest candidate to generate electricity from dilute waste heat sources.

Good thermoelectric materials are indispensable to thermoelectric power generation. However, these materials should have not only a high dimensionless thermoelectric figure of merit $ZT$ as defined in Eq. 1, but also high chemical stability and should not contain harmful elements.

$$ZT = S^2 T / \kappa \rho \qquad (1)$$

Here, $S$ is the Seebeck coefficient, $T$ the absolute temperature, $\kappa$ the thermal conductivity, and $\rho$ the electrical resistivity.

Temperature range of waste heat is spread from 350 K to 1200 K or higher. Almost all metallic compounds have a problem of oxidation at such temperatures. Oxide thermoelectric materials are considered to be promising ones because of their durability against high temperature, cost, no content of toxic elements, and so on. Especially, some layered $CoO_2$ compounds show high $ZT$ values in air [2-4]. Fabrication and properties of modules using oxide materials have been produced [5-8]. Though $ZT$ values around unity are obtained at temperatures higher than 800 K, enhancement of $ZT$ is necessary at temperatures lower than 800 K for waste heat application. It is not efficient that only single oxide devices are used over the wide temperature range. Whereat, cascade modules accumulated oxide and $Bi_2Te_3$ modules have been developed to obtain high generating power over wide temperature range. Actually, cascading is effective to enhance power density at hot side temperature higher than 873 K [9]. However, $ZT$ values of oxide materials are good at 800 K or higher and common $Bi_2Te_3$ modules can be used at temperatures lower than 473 K or lower because of degradation of junctions between thermoelectric legs and electrodes. In other words, there is a "blank area" at middle temperature range (473-800 K). Development of not only thermoelectric materials with high $ZT$ values and oxidation resistance, but also modules with high generating power density and durability are important tasks of pressing urgency. Silicide compounds are one of the strongest candidates for the middle temperature range because of formation of passive layer on the surface. In fact, $MnSi_{1.75}$ and $FeSi_2$ are well-known as thermoelectric materials usable at high temperature even in air [10-13]. Thermoelectric properties of an n-type silicide material are discussed in this paper.

**EXPERIMENT**
Preparation

Precursor ingots were prepared by arc melting Si, Mn, Ti, V, Cr, Fe, Co, Ni, Cu, and Al metallic chips with an atomic ratio of Mn : M : Si : Al = 3-x : x : 4 : 3, where x is from 0 to 1.0 and M indicates Ti, V, Cr, Fe, Co, Ni, Cu. After polishing the black surface of the ingots to remove oxides formed during melting, the ingots were ground using an agate mortar and pestle in air. The powder was packed into disk-shaped pellets. The precursor pellets were sintered by pulse electric current sintering (SPS) at 1023 K for 15 min under a uniaxial pressure of 30 MPa using carbon die in vacuum.

Measurement

X-ray diffraction (XRD) of the powder samples after SPS was used to investigate the crystallographic structure and purity. The XRD patterns were analyzed using the Rietveld method with the help of the Jana 2006 software to calculate cell parameters. Microstructure of the sintered samples was observed using a scanning electron microscope (SEM) in back scattering mode. The analysis of elemental composition was performed by energy dispersed X-ray analysis (EDX). Transmission electron microscope (TEM) observation and electron diffraction (ED) measurement were carried out to assign the crystallographic structure.

The samples were cut into rectangular bars, about 3 x 3 mm square in cross-section and 5-10 mm long. The direction of temperature difference for the Seebeck coefficient ($S$) measurement and of current flow for electrical resistivity ($\rho$) measurement is perpendicular to the pressing axis of the SPS processing. The Seebeck coefficient values were calculated from a plot of thermoelectric voltage against the temperature differential as measured at 373-973 K in air using an instrument designed by our laboratory. Two Pt-Pt/Rh (R-type) thermocouples were adhered to both ends of the samples using silver paste. The Pt wires of the thermocouples were used for voltage terminals. Measured thermoelectric voltage was plotted against the temperature difference. The slope corresponds to the apparatus Seebeck coefficient. The actual $S$ values were obtained by correction using the $S$ values of the Pt wire. Electrical resistivity was measured using the standard DC four-probe method in air from room temperature to about 900 K. Silver paste was used for the connections between the samples and both the current and voltage lead wires. Thermal conductivity ($\kappa$) measurement was carried out using the laser flash method in a direction parallel to the pressing axis during SPS processing.

**RESULTS and DISCUSSION**

<u>Chemical composition and crystallographic structure</u>

The XRD pattern for the sample with a starting composition of Mn : Si : Al = 3 : 4 : 3 is shown in Fig. 1. The crystallographic structure of all the samples can be assigned to a hexagonal closed-packed $CrSi_2$ structure [14]. The $a$- and $c$- cell parameters for the Mn : Si : Al = 3 : 4 : 3 sample are 0.447 nm and 0.644 nm, respectively. Some peaks due to secondary phases are observed. The secondary phases can be detected by microscopic observation using SEM. The element ratio measured by EDX are Mn : Si : Al = 3.0 : 4.0 : 2.3 (bright matrix portion) and Mn : Si : Al = 1.0 : 2.9 : 2.1 (dark portion) (Fig. 2). The bright portion corresponds to the main phase. The elemental composition of the main phase disagrees with the starting one. The decrease in Al seems to be caused by oxidation during arc melting. The oxides are excluded to the surface of the ingots and removed by polishing. As a result, the composition of the main phase should be close to Mn : Si : Al = 3.0 : 4.0 : 2.0. This result implies that Mn occupies the Cr-site, and Si and Al the Si-site of the $CrSi_2$ structure. Hereafter, the main phase is indicated as 342 phase. Though the secondary phases exist in the samples, the amount of them is small and effects on thermoelectric properties seem to be negligible.

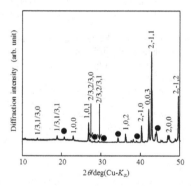

**Figure 1.** X-ray diffraction pattern for the sample with $Mn_3Si_4Al_3$ of the starting composition after pulse electric current sintering (SPS). Closed circles and arrows correspond to the diffraction peaks due to secondary phases and superlattice structure, respectively.

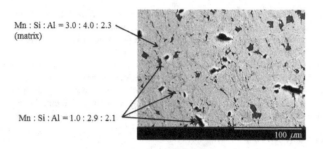

Mn : Si : Al = 3.0 : 4.0 : 2.3 (matrix)

Mn : Si : Al = 1.0 : 2.9 : 2.1

100 μm

**Figure 2.** Scanning electron microscopic image for the sample with $Mn_3Si_4Al_3$ of the starting composition after pulse electric current sintering (SPS).

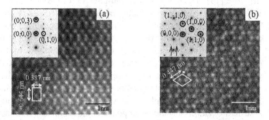

**Figure 3.** Transmission electron microscopic images and electron diffraction patterns for the $Mn_3Si_4Al_2$ phase for the direction of incidence of [1 0 0] (a) and [0 0 1] (b). Arrows in the electron diffraction patterns indicate the existence of a superlattice structure.

More detail structure is also investigated by TEM and ED measurement from some incident directions (Fig. 3). Strong fundamental diffraction spots observed were confirmed to be originated from the $CrSi_2$ structure as indexed in Fig. 3. In addition, our careful analysis revealed the presence of weak diffraction spots as indicated by arrows. These spots were present at (1/3, 1/3, 0) positions in the reciprocal space, which suggest a superlattice structure. The diffraction peaks indexed by (1/3, 1/3, 0), (1/3, 1/3, 1), (2/3, 2/3, 0), and (2/3, 2/3, 1) are also observed in XRD pattern in Fig. 1. This means that there is a structural modulation with a three-fold period in the [1 1 0] direction with respect to the $CrSi_2$ structure. The relationship between the superlattice structure and thermoelectric or other properties is an open question.

**Optimization of substituting element**

Temperature dependence of Seebeck coefficient $S$ and electrical resistivity $\rho$ for the $Mn_3Si_4Al_2$ sample in Fig. 4 (a). $S$ values increase up to 773 K, then suddenly decrease at 973 K. Maximum value of $S$ is 93 $\mu$V/K at 773 K. The $\rho$ value increases with increasing temperature up to about 800 K showing a metallic behavior. This character changes to an insulator-like one around 800 K. This transition temperature almost corresponds to the temperature of the sudden

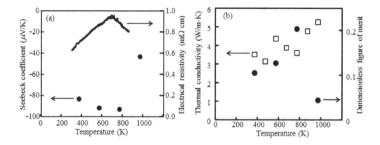

**Figure 4.** Temperature dependence of Seebeck coefficient and electrical resistivity (a) and thermal conductivity and dimensionless figure of merit $ZT$ (b) for the $Mn_3Si_4Al_2$ of the starting composition after SPS.

decrease in $S$. Considering these behaviors of $S$ and $\rho$, the 342 phase is in the intrinsic range at temperatures higher than 800 K. Figure 4 (b) shows the temperature dependence of the thermal conductivity $\kappa$. Basically the $\kappa$ values increase with increasing temperature and reach as high as 5 W/m-K. In order to obtain high $ZT$, the $\kappa$ values should be reduced. Though $ZT$ for the non-substituted 342 sample increases with increasing temperature up to 773 K, it decreases suddenly at 973 K. This is due to the decrease in Seebeck coefficient because of transition to the intrinsic range. The maximum $ZT$ reaches 0.2 at 773 K. In order to enhance the $ZT$ values, controlling the career density by elemental substitution for elevation of the transition temperature into the intrinsic range and optimizing the microstructure including the secondary phases as scattering sites of phonons for low $\kappa$ values are vital.

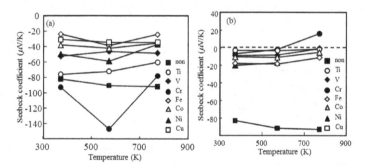

**Figure 5.** Temperature dependence of Seebeck coefficient for the samples with $Mn_3Si_4Al_3$ (non), $Mn_{2.7}M_{0.3}Si_4Al_2$ (a) and $Mn_2MSi_4Al_2$ (b). M is Ti, V, Cr, Fe, Co, Ni, or Cu.

107

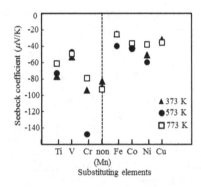

**Figure 6.** Seebeck coefficient of $Mn_3Si_4Al_2$ (non) and $Mn_{2.7}M_{0.3}Si_4Al_2$ samples at 373-773 K.

The elemental substitution has been carried out at Mn-site with 3d transition metals. Figure 5 indicates temperature dependence of Seebeck coefficient for the substituted 342 samples. Though the starting composition was $Mn_{3-x}M_xSi_4Al_3$ as mentioned above, the final compositions for the matrix main phase of all the samples were close to $(Mn+M) : Si : Al= 3 : 4 : 2$. Outstanding increase in $S$ values is observed only for the Cr-substituted sample with x = 0.3. The $S$ values decrease in all other samples with both x = 0.3 and 1.0. The reduction of the $S$ values tends to be small in the samples substituted with the prior elements compared with the posterior ones than Mn in the periodic table (Fig. 6). This result indicates that not only carrier density but ionic radii should affect on the $S$ values.

Temperature dependence of $\rho$ and power factor (= $S^2/\rho$) for the samples substituted by x = 0.3 are shown in Fig. 7. The $\rho$ values tend to be lower in the samples substituted by the posterior elements than Mn because of high level of electron doping. This result is consistent

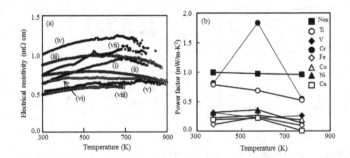

**Figure 7.** Temperature dependence of electrical resistivity (a) and power factor (= $S^2/\rho$) for the samples of $Mn_3Si_4Al_2$ (i) and $Mn_{2.7}M_{0.3}Si_4Al_2$. M : Ti (ii), V (iii), Cr (iv), Fe (v), Co (vi), Ni (vii), Cu (viii).

108

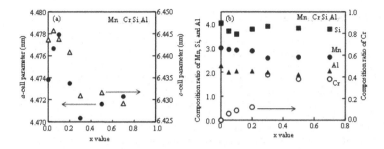

**Figure 8.** *a*- and *c*-cell parameters (a) and composition ratio of each element (b) for the Mn$_{3-x}$Cr$_x$Si$_4$Al$_2$.

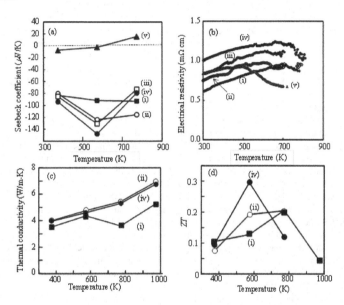

**Figure 9.** Temperature dependence of Seebeck coefficient (a), electrical resistivity (b), thermal conductivity (c), and dimensionless figure of merit *ZT* (d) for the Mn$_{3-x}$Cr$_x$Si$_4$Al$_3$ samples. x = 0 (i), 0.05 (ii), 0.1 (iii), 0.3 (iv), and 1.0 (v).

with the low *S* values. The substitution with Cr is effective to enhance power factor. In order to optimize the Cr composition, the 342 samples with different x values of Mn$_{3-x}$Cr$_x$Si$_4$Al$_2$ have been prepared.

## Thermoelectric property of $Mn_{3-x}Cr_xSi_4Al_2$

Figure 8 (a) indicates that both $a$- and $c$- cell parameters obtained from the XRD patterns increase up to x = 0.1 and 0.05, respectively and decrease with increasing Cr-substitution. The decrease in the parameters become almost constant at x higher than 0.3. According to Vegard's law, the lattice parameters should be increased by substitution of Mn with Cr, which tends to have larger ionic radii than Mn in the case of the same coordination number valence, and so on [15]. The reason of the shrinkage of the lattice is an open question and should be investigated in more detail with consideration for the coordination number, valence, and so on. The $a$- and $c$-cell parameters are independent of the starting composition of Cr in x ≥ 0.3. This is due to the constant content of Cr in the 342 phase. As mentioned below, the Cr x = 0.3 is near solubility limit of Cr into the Mn-site, This can be more clearly observed by EDX analysis. The composition ratios for each element are plotted in Fig. 8 (b). The ratio of Cr increases steadily up to x = 0.2, then discontinuously jumps up and is saturated in x > 0.3. The composition ratios of Si and Al are almost independent of the x values in the Cr-substituted samples.

Temperature dependence of $S$ is shown in Fig. 9 (a). The $S$ values for all samples increase up to 573 K in 0 < x < 0.3, then suddenly decrease at 773 K in the 0.05 < x < 0.3. This result indicates the transition temperature into the intrinsic range drops by the Cr-substitution. The $S$ values decrease with increasing temperature from 373 K in the x = 0.5 and 0.7 samples. The Cr-substitution up to x = 0.3 enhances the $S$ values at 573 K. Maximum value of $S$ reaches a high 160 $\mu$V/K at x = 0.3, which is more than 1.6 times greater than that for the non Cr-substituted sample.

It is clear that the $\rho$ values are elevated by the Cr-substitution (Fig. 9 (b)). Electrical conduction behavior of all samples changes from a metal-like temperature dependence to an insulator-like one at 623-723 K. These transition temperatures almost correspond to the temperature at which the decrease in $S$ begins. The Cr-substitution seems to decrease the carrier density in the 342 phase at temperatures lower than 573 K. Of course the secondary phases and microstructure should be considered as the reasons for the change in electrical properties. The transition temperature into the intrinsic range is lowered by the Cr substitution.

On the other hand, the $\kappa$ values increase with increasing temperature in all the samples (Fig. 9 (c)). The Cr-substitution pushes the $\kappa$ values up slightly, though $\rho$ is elevated, namely thermal conduction due to electrons is suppressed. From XRD and EDX analysis, the amount of secondary phases is decreased by Cr-substitution [16]. Though the secondary phases may act as phonon scattering sites, the changes in crystallographic structure or other microstructure should be paid attention. Moreover, the decrease in $\kappa$ at 773K for the non-substituted sample has to be discussed in viewpoints of the crystallographic structure, secondary phases, transition to the intrinsic range and so on hereafter.

Though the $ZT$ values for the samples with 0 ≤ x ≤ 0.05 increase with increasing temperature up to 773 K, the Cr-substituted samples have a peak at 573 K (Fig. 9 (d)). $ZT$ is enhanced up to 0.3 at 573 K by the Cr-substitution with x = 0.3. In order to increase the $ZT$ values, reduction of the $\kappa$ values is indispensable. Optimizing the microstructure, such as inclusions of secondary phases should be necessary to enhance scattering of phonon.

Figure 10 (a) shows the $\rho$ values measured continuously at 873 K in air. Though the measurement was performed for 48 h, the $\rho$ values were maintained constant. Additional diffraction peaks which can be assigned to alumina, silica, or complex oxides of the constituent elements are detected in the XRD patterns for the heated sample (Fig. 10 (b)). The color of the surface of the sample after heat treatment at 873 K turns from silvery to brown color (Fig. 11).

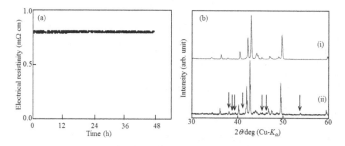

**Figure 10.** Electrical resistivity measured continuously at 873 K in air (a) and X-ray diffraction patterns before (i) and after (ii) heat treatment at 873 K for 48 h in air (b) for the $Mn_3Si_4Al_2$ sample.

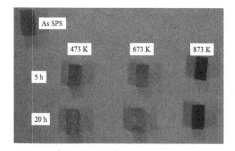

**Figure 11.** Photograph of $Mn_3Si_4Al_2$ samples after SPS processing and heat treatment in 473–873 K for 5 or 20 h in air.

The brown layer can be removed easily by polishing with a sheet of emery paper and the silvery surface appears under the brown layer. These results indicate that the 342 phase is oxidized, but it occurs only the surface. Because the formed oxide layer acts as a passive layer to protect the inside of the 342 bulk from oxidation, the $\rho$ values are kept constant at high temperatures even in air. In order to make the 342 material into a good thermoelectric device in the middle temperature range, high sintered density is necessary to prevent oxygen penetration into the inside of the bulk.

A thermoelectric module has been already prepared using the p- and n-type devices composed of $MnSi_{1.75}$ and $Mn_3Si_4Al_2$, respectively [16]. The maximum output power reached 9.4 W corresponding to 2.3 kW/m² against the surface area of the substrate, at the hot temperature of 873 K.

## CONCLUSIONS

Thermoelectric properties of $Mn_3Si_4Al_2$ (342 phase) with a $CrSi_2$ structure were investigated. This material shows an n-type character and $ZT$ of 0.2 at 773 K. The element substitution of Mn with the 3d transition metals is succeeded. Only the Cr-substitution is effective to improve $ZT$ values, which reach 0.3 at 573 K in the $Mn_{2.7}Cr_{0.3}Si_4Al_2$. Since the 342 phase is into the intrinsic range at temperatures higher than 573-773 K, $ZT$ values markedly decrease. Electrical resistivity measured at 873 K was constant for more than two days in air. This indicates that the 342 devices have good oxidation resistance at high temperature in air. This is caused by the formation of passive oxide layers around the surface.

## REFERENCES

1. World Energy Outlook, 2009 Edition, International World Energy, 2009.
2. I. Terasaki, Y. Sasago and K. Uchinokura, Phys. Rev. B 56, R12685 (1997).
3. R. Funahashi, I. Matsubara, H. Ikuta, T. Takeuchi, U. Mizutani and S. Sodeoka, Jpn. J. Appl. Phys., 39, L1127 (2000).
4. R. Funahashi and M. Shikano, Appl. Phys. Lett., 81, 1459 (2002).
5. R. Funahashi, M. Mikami, T. Mihara, S. Urata, N. Ando, J. Appl. Phys., 99, 066117 (2006).
6. S. Urata, R. Funahashi, T. Mihara, A. Kosuga, S. Sodeoka, T. Tanaka, Int. J. Appl. Ceram. Tech., 4, 535 (2007).
7. P. Tomes, C. Suter, M. Trottmann, A. Steinfeld, A. Weidenkaff, J. Mater. Res. 26, 1975 (2011).
8. A. Inagoya, D. Sawaki, Y. Horiuchi, S. Urata, R. Funahashi, I. Terasaki, J. Appl. Phys., 110, 123712 (2011).
9. R. Funahashi, Sci. Adv. Mater., 3, 683 (2011).
10. T. Yamada, Y. Miyazaki, H. Yamane, Thin Solid Films, 519, 8524 (2011).
11. I. Aoyama, M. I. Fedorov, V. K. Zaitsev, F. Y. Solomkin, I. S. Eremin, A. Y. Samunin, M. Mukoujima, S. Sano, and T. Tsuji, Jpn. J. Appl. Phys., 44, 8562 (2005).
12. R. Wolfe, J. H. Wernick, and S. E. Haszko, Phys. Lett., 19, 449 (1965).
13. K. Morikawa, H. Chikauchi, H. Mizoguchi, S. Sugihara, Mater. Trans., 48, 2100 (2007).
14. JCPDS Card No. 35-0781.
15. R. D. Shannon Acta Cryst., A32, 751 (1976).
16. R. Funahashi, Y. Matsumura, H. Tanaka, T. Takeuchi, W. Norimatsu, E. Combe, R. O. Suzuki, Y. Wang, C. Wan, S. Katsuyama, and K. Koumoto, J. Appl. Phys, 112, 073713 (2012).

**Nanocomposites and Nanostructured Materials**

Mater. Res. Soc. Symp. Proc. Vol. 1490 © 2012 Materials Research Society
DOI: 10.1557/opl.2012.1642

# Thermoelectric properties of Bi-FeSb₂ nanocomposites: Evidence for phonon-drag effect

Mani Pokharel, Machhindra Koirala, Huaizhou Zhao, Kevin Lukas, Zhifeng Ren, and Cyril Opeil

*Department of Physics, Boston College, Chestnut Hill MA 02467*

## Abstract

The thermoelectric properties of Bi-FeSb₂ nanocomposites are reported. The electrical resistivity and the Seebeck coefficient measurements show a significant dependence on bismuth concentration. Our results reveal that the shifting of the Seebeck peak in FeSb₂ nanocomposites is purely a grain size-effect. The thermal conductivity data indicates a presence of an electron-phonon interaction. Over all, our analysis of the the thermoelectric properties of Bi-FeSb₂ nanocomposites provide additional evidence for phonon-drag in FeSb₂.

## Introduction

FeSb₂ has drawn extensive research efforts in the recent years [1-7] because of its colossal Seebeck coefficient of -45000 $\mu VK^{-1}$ at ~ 10 K, which makes this compound a potential candidate for thermoelectric cooling applications. Despite the huge power factor (PF) of 2300 $\mu WK^{-2}cm^{-1}$ at ~ 10K, the $ZT$ values for single crystals are limited by the large value of the thermal conductivity. In our earlier work [6, 7], we successfully reduced the thermal conductivity by three orders of magnitude using a nanostructuring approach. However due to the suppressed peak value of the Seebeck coefficient, $ZT$ values for nanocomposite samples were still far too low for use in practical applications.

In general, the Seebeck coefficient is the sum of two independent contributions.

$$S = S_{diffusion} + S_{phonon} \tag{1}$$

The diffusion term arises due to the spontaneous diffusion of carriers in the presence of thermal gradient whereas the phonon term is associated with the preferential scattering of carriers by phonons from the hot to the cold end, also called the phonon-drag effect. The origin of the huge Seebeck coefficient in FeSb₂ is not completely understood yet. A huge electron diffusion caused by a strong electron-electron correlation has been suggested by many authors [1-5] as a possible cause. Other authors [8, 9], however, emphasized the non-electronic origin and suggested phonon-drag as a possible mechanism. In this paper, we study the thermoelectric properties of Bi-FeSb$_2$ nanocomposites. The preliminary results provide additional evidence for phonon-drag effects in FeSb$_2$.

## Experiment and Methods

FeSb$_2$ was synthesized by ingot formation through melting and solidification inside an evacuated quartz tube. The ingot was then ball-milled for 15 hours to obtain a fine powder. Bi-powder in the amount of 2, 6 and 10 % by weight was added to the FeSb$_2$ powder the mixture was ball-milled for additional 6 hours. The final powder was then hot pressed at 200 °C for 2 minutes. A pressure of about 70 MPa was applied during pressing. The electrical resistivity ($\rho$), Seebeck Coefficient ($S$), and thermal conductivity ($\kappa$) of the samples were all measured on a Physical Property Measurement System (PPMS) from Quantum Design.

## Results and Discussion

Figure (1) shows the temperature dependence of the electrical resistivity for the four samples. All the samples exhibit semiconducting behavior with a negative temperature coefficient. With the increasing percentage of bismuth, the electrical resistivity decreases throughout the temperature range of 5-300 K. The decreasing electrical resistivity is attributed to the increased carrier concentration from the addition of bismuth. Greater difference in electrical resistivity at low temperature might be is attributed to the semi metallic nature of bismuth.

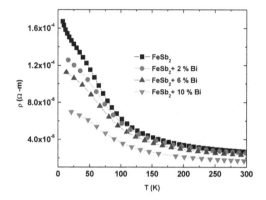

**Figure (1):** Electrical resistivity as a function of temperature for the four nanocomposite samples.

Figure (2) shows the temperature dependence of the Seebeck coefficient from 5 to 300 K for the samples. Qualitatively, all the samples exhibit a similar temperature dependence transitioning from p-type to n-type at ~ 165 K. The absolute value of the peak in the Seebeck coefficient decreases with increasing bismuth content. This result is expected because the increase in carrier concentration decreases Seebeck coefficient. In our earlier work [6], the Seebeck peaks shifted to higher temperature with decreasing hot pressing temperature i.e. decreasing grain size. In contrast, peaks in Seebeck coefficient occur at roughly the same temperature, ~ 45 K, for all samples used in this study. Since all the samples were hot pressed at the same temperature (200 °C) and hence should have similar grain sizes, this observation further confirms that the shifting in Seebeck peak is purely a grain-size effect. Here we note that such a size-dependent shift of the Seebeck peak has been shown to be one of the striking features of the phonon-drag dominated systems [10-12].

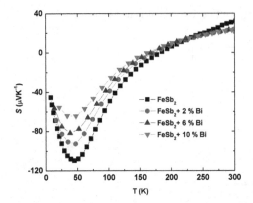

**Figure (2):** Seebeck Coefficient as a function of temperature for the four nanocomposite samples.

Figure (3) shows the temperature dependence of the thermal conductivity. The thermal conductivity values for all the samples are drastically reduced compared to the single crystal samples. Under the approximation that the lattice and carriers contribute independently, the total thermal conductivity is given by,

$$\kappa_{total} = \kappa_{lattice} + \kappa_{carrier} \tag{2}$$

Based on this approximation, one would expect increased total thermal conductivity in Bi-FeSb$_2$ nanocomposites due to greater contributions from carriers. However it is surprising that the Bi-FeSb$_2$ nanocomposites exhibit smaller thermal conductivity compared to the pure FeSb$_2$ nanocomposite violating the independent approximation. While the decrease in the thermal conductivity could come from decreased lattice thermal conductivity due to point defect scattering, electron-phonon interaction could also be another possible mechanism responsible for the violation of independent approximation as reported in FeSi by Sales *et al.* [13].

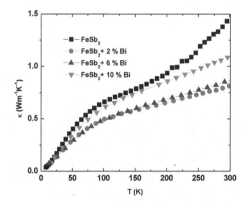

**Figure (3):** Thermal conductivity as a function of temperature for the four nanocomposite samples.

Much stronger evidence for the phonon-drag effect has been seen in our earlier work where the thermal conductivity and the Seebeck coefficient were found to have strong grain-size dependence [14]. The analysis there showed a significant contribution from phonon-drag to the large Seebeck peak for coarse-grained samples. Our present work provides further evidence of the phonon-drag effect. A complete study would provide more clear information along with improved $ZT$ of $FeSb_2$ samples. This further work is underway.

## Conclusion

The shifting of the Seebeck peak of $FeSb_2$ is purely a size-effect as seen in many other phonon-drag dominated systems. The independent approximation for thermal conductivity does not seem to work for $FeSb_2$ which could be attributed to the electron-phonon coupling. In conclusion, the thermoelectric properties of $Bi$-$FeSb_2$ nanocomposites support the presence of phonon-drag effects in $FeSb_2$.

## References

1. A. Bentien, S. Johnson,G.K.H. Madsen,B.B Iversen and F. Steglich *EPL*, 80(**2007**) 17008
2. P. Sun, N. Oeschler, S. Johnsen, B. B. Iversen, and F. Steglich,*Dalton Trans.*, **2010**,39,1012-1019

3. P. Sun, N. Oeschler, S. Johnsen, B. B. Iversen, F. Steglich, *Phys. Rev. B.* **2009**, 79 (15), 153308.
4. A. Bentien, G. K. H. Madsen, S. Johnson, B. B. Iversen, *Phys. Rev. B.* **2006**, 74 (20), 205105.
5. P. Sun, M. Søndergaard, Y. Sun, S. Johnsen, B. B. Iversen, F. Steglich, *Appl. Phys. Lett.* **2011**, 98, 072105.
6. H. Zhao, M. Pokharel, G. Zhu, S. Chen, K. Lukas, Q. Jie, C. Opeil, G. Chen, and Z. Ren, *Appl. Phys. Lett.* 99, 163101 **(2011)**
7. M. Pokharel, H. Zhao, K. Lukas, Z. Ren, and C. Opeil, *MRS Proceedings/Volume* 1456 *Spring* **2012**
8. H. Takahashi, R. Okazaki, Y. Yasui, I. Terasaki, *Phys. Rev. B* **2011**, *84*, 205215.
9. J. M.Tomczak, K. Haule, T. Miyake, A. Georges, G. Kotliar, *Phys. Rev. B* **2010**, *82*, 085104.
10. Q. R. Hou, B. F. Gu, Y. B. Chen, Y. J. He, *Modern Physics Letters B*, **2011**, *25*, 1829.
11. J. P. Issi, J. Boxus, *Cryogenics* **1979**, *19*, 517.
12. T. H. Geballe, and G.W. Hull, *Phys. Rev.* **1954**, *94*, 1134.
13. B.C. Sales, O. Delaire, M. A. McGurie, and A. F. May, *Phys. Rev. B* **2011**, 83, 125209
14. Mani Pokharel, Huaizhou Zhao, Kevin Lukas, Bogdan Mihaila, Zhifeng Ren, and Cyril Opeil, *arXiv*:1210.2999 [cond-mat.mes-hall]

Mater. Res. Soc. Symp. Proc. Vol. 1490 © 2012 Materials Research Society
DOI: 10.1557/opl.2012.1643

Fabrication of nanostructured bulk Cobalt Antimonide (CoSb₃) based skutterudites via bottom-up synthesis

M. Saleemi*,1, M. Y. Tafti[1], M. S. Toprak[1], M. Stingaciu[2], M. Johnsson[2], M. Jägle[3], A. Jacquot[3], M. Muhammed[1]

[1] Department of Materials and Nanophysics, KTH Royal Institute of Technology, Kista-Stockholm, Sweden.
[2] Department of Materials and Environmental Chemistry, Stockholm University, Stockholm, Sweden.
[3] Fraunhofer-Institut für Physikalische Messtechnik IPM, 79110 Freiburg, Germany

Keywords: Bottom up synthesis, Thermoelectric, Nanostructured, Skutterudite, Cobalt Antimonide, Spark Plasma Sintering.

ABSTRACT

Skutterudites are known to be efficient thermoelectric (TE) materials in the temperature range from 600 K to 900 K. Dimensionless figure of merit (ZT) for filled skutterudite TE materials have been reported as ca. 1 at 800 K. Novel nano- engineering approaches and filling of the skutterudites crystal can further improve the transport properties and ultimately the ZT. Although classified among the promising TE materials, research on their large-scale production via bottom up synthetic routes is rather limited. In this work, large quantity of cobalt antimonide (CoSb₃) based skutterudites nanopowder (NP) was fabricated through a room temperature co-precipitation precursor method. Dried precipitates were process by thermo-chemical treatment steps including calcination (in air) and reduction (in hydrogen). CoSb₃ NPs were then mixed with silver (Ag) nanoparticles at different weight percentages (1%, 5% and 10% by wt) to form nanocomposites. Skutterudite NP was then consolidated by Spark Plasma Sintering (SPS) technique to produce highly dense compacts while maintaining the nanostructure. Temperature dependent TE characteristics of SPS'd CoSb₃ and Ag containing nanocomposite samples were evaluated for transport properties, including thermal conductivity, electrical conductivity and Seebeck coefficient over the temperature range of 300 - 900 K. Physicochemical, structural and microstructural evaluation results are presented in detail.

INTRODUCTION

Skutterudites have been widely investigated as potential next generation thermoelectric (TE) materials for energy harvesting. Due to their unique crystal structure and exclusive electrical transport properties, they are known to be one of the promising TE material in the temperature range of 600 to 900 K.[1] Bulk productions of skutterudite based TE materials with high purity and homogeneity is major obstacle to commercialize the skutterudite TE modules. Furthermore, relatively high thermal conductivity causes the poor TE performance in such materials. Cobalt anitmonide (CoSb₃) based skutterudites have attained great attention because of their low volatility, abundancy of the elements involved and lower production cost than other classes of skutterudites.[2-3] CoSb₃ is a narrow band semiconductor material with excellent electrical transport properties among skutterudites. CoSb₃ represents cubic crystal and belongs to space group $IM_3$. In this structure, cobalt (Co) atoms form a cubic crystal with four-membered antimony (Sb) rings occupying six of the eight cubic sub-cages, while the two remaining cages are empty. [3] Many reports have reported on favorable effect on figure of merit (ZT) of CoSb$_3$ base materials by utilizing the nanostructures in conjunction with grain boundary pinning.[4] Nano-engineering of CoSb$_3$ compounds can be one approach to resolve the TE efficiency problems, as discussed in the literature. Adding the nanostructures into the CoSb$_3$ by filling the voids or producing the grain boundaries inclusions may reduce the thermal conductivity and further enhances the TE performance.[5] These mechanisms directly increase the phonon scattering by introducing filled atoms in the crystal structure and grain boundary inclusions.[6] In the present work, we report an investigation about grain boundary inclusions on CoSb$_3$ nanostructures and their TE performances. CoSb$_3$ was fabricated using bottom-up chemical synthesis and the effect of silver (Ag) nanoparticles was studied as a function of embedded nanoparticles concentration. Nano Ag-CoSb$_3$ composites were prepared and the phase purity, microstructural analysis and the TE properties were evaluated.

## EXPERIMENT

All the chemicals were purchased from Sigma Aldrich with 99.999 % purity. CoSb$_3$ was synthesized via bottom up chemical alloying route as introduced and developed in our research group.[7-9] In this method, metal ions precursors were mixed under controlled thermodynamic conditions and metal oxide precipitates were achieved at room temperature. These precipitates were filtered off, washed with de-ionized water several times and dried at 60 °C under vacuum overnight. Subsequent thermal processing such as calcination at 350 °C in air and reduction at 500 °C in hydrogen environment formed the final product. Further details about the synthesis of CoSb$_3$ nanopowder can be found elsewhere.[7] Commercially available nano-silver (Ag) with average particle size of 50 nm were utilized as nanoinclusions. To produce nanocomposites three different weight percentages (1%, 5% and 10% by wt) of Ag NP were dispersed by mechanical mixing and water bath sonication in wet medium. The solvent was then evaporated at low temperature in drying oven under vacuum.

### Characterization and Processing

Dried nanopowder contains the mixture of CoSb$_3$ and Ag; it was confirmed by performing scanning electron microscopy (SEM) and energy dispersive spectroscopy (EDS) by utilizing Gemini Zeis FEG-SEM system. X ray diffraction (XRD) analysis was performed to investigate the crystal phases and calculation of crystallite size. Spark plasma sintering (SPS) from Dr. Sinter 2000, was used to compact the nanopowder into pellet. To attain a high compaction density and preserve nanostructure, we have optimized SPS consolidation parameters in our previous reports.[10] In this work we have utilized the optimal conditions for CoSb$_3$ based material. Nanopowder was sintered at 500 °C with heating rate 50 °C/min, 2 minutes holding and applied pressure of 75 MPa. The density of compacted samples was measured by Archimedes principle and SEM analysis was performed on the fractured surface to investigate the dispersion of Ag nanoparticles in CoSb$_3$ phase. TE evaluations were performed by using in house build ZT meter by Fraunhofer-IPM.[11] Compacted samples were cut into specific dimensions according to the sample holder requirements, Seebeck coefficient (S), Electrical conductivity (σ), thermal conductivity (κ) was measured simultaneously up to 825 K.

**RESULTS and DISCUSSION**

As synthesized CoSb$_3$ nanopowder and as prepared Ag-CoSb$_3$ nanocomposites were subjected to XRD analysis for the crystal structure identification. Figure 1 (a) shows the XRD pattern from as synthesized CoSb$_3$ nanoparticles produced via chemical alloying method. It presents a single phase of CoSb$_3$ skutterudite and all the reflected peaks were indexed with the JCPDS reference file no:076-0470. Figure 1 (b-d) represents XRD patterns from the Ag-CoSb$_3$ nanocomposites, clearly revealing the presence of CoSb$_3$ and Ag. The intensity of Ag peaks increase from figure 1(b) to 1(d) in agreement with the increasing content of Ag NP. All the peaks from Ag were indexed with JCPDS reference file no:04-0783.

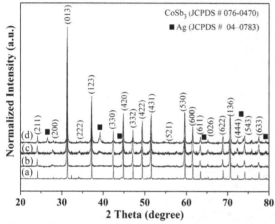

**Figure 1.** XRD Patterns of skutterudite nanostructures; (a) As synthesized CoSb$_3$, (b) CbS_Ag1, (c) CbS_Ag5, and (d) CbS_Ag10.

In table 1, describes the sample details, concentration of nanoinclusions and material composition, followed by the achieved SPS compaction densities with average grain size in each sample. Optimized SPS conditions have provided high compaction densities (*more than 95 %*) and preserved nanostructures, which may influence the TE properties.

**Table 1.** Sample details and compaction densities

| Sample IDs | Ag NP Concentration (Wt %) | Material Composition (Molecular formula) | SPS Compaction Density (%) | Grain size (nm) |
|---|---|---|---|---|
| CbS_Ag1 | 1 | CoSb$_3$Ag$_{0.04}$ | 96 | ~ 250-300 |
| CbS_Ag5 | 5 | CoSb$_3$Ag$_{0.2}$ | 96 | ~ 300 |
| CbS_Ag10 | 10 | CoSb$_3$Ag$_{0.4}$ | 95 | ~ 300-350 |

SEM micrograph in Figure 2 (a) representes the as prepared Ag-CoSb$_3$ nanocomposite. Two different sizes of particles were observed, one is in the range of 50 nm (Ag NP) which are homogeneously dispersed and second ones are in the range of 200 nm which represents as synthesized CoSb$_3$ nanostructures. Figure 2(b-d) are SEM micrographs from the cleaved surfaces of SPS compacted composite samples. Ag nanoparticles (small particles) are observed to be homogenously dispersed at the grain boundaries, which may provide the effect of nanoinclusions or grain boundary pinning. With the increase in concentration of Ag, more accumulation at the grain boundaries are observed. These two material phases were confirmed by energy dispersive spectroscopy (EDS), confirming the smaller particles being Ag and larger grains as CoSb$_3$.

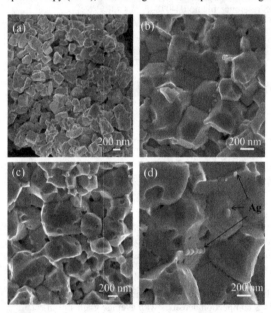

**Figure 2.** SEM micrographs; (a) Bulk nanostructures of CoSb$_3$ dispersed with Ag Nanoparticles, (b) SPS'ed CbS_Ag1, (c) SPS'ed CbS_Ag5, (d) SPS'ed CbS_Ag10.

## Thermoelectric Evaluation

SPS compacted sample was cut with the specific dimensions to perform the TE characterization. In house built ZT Meter (at Fh-IPM) was used in the temperature range of 325K to 825K. Seebeck coefficient (S), electrical conductivity ($\sigma$), thermal conductivity ($\kappa$) and TE figure of merit ($ZT$) are presented in figure 3 (a-d), respectively. Figure 3(a), reveals p-type semiconductor behavior of high content Ag-CoSb$_3$ nanocomposites and Seebeck values reached to maximum, 60 $\mu$V/K at 600 K for 10 wt% Ag content. Ag1 sample starts with n-type characteristic that shifts to p-type around 470 K. This value is lower than the other metal doped

systems but comparable to the metal oxide CoSb₃ composites.[12] In figure 3(b), electrical
conductivity have shown varying results from three different Ag concentration of CoSb₃ based
skutterudites. With 1 and 5 wt% of Ag content, the σ is very low as compared to other reports on
doped CoSb₃ systems but 10 wt% Ag has shown 4 times improvement in electrical conductivity.
Thermal conductivity of prepared nanocomposites were measured and presented in figure 3(c), it
shows that CbS_Ag5 have minimum thermal conductivity value, about 2 W/mK at 600K, but it
does not increase the overall ZT as the electrical conductivity is not promising. Figure 3(d) shows
that CbS_Ag10 have 3 times higher ZT than lower concentration of Ag doped composites which
is mainly due to the contribution of higher electrical conductivity.

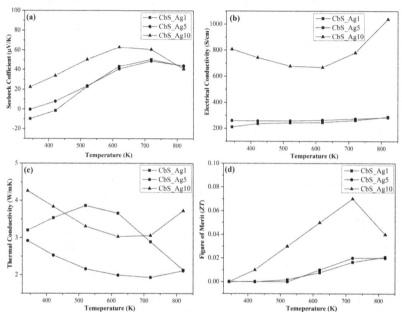

**Figure 3.** TE characterizations of Ag-CoSb₃ Nanocomposites; (a) Seebeck Coefficient, (b)
Electrical Conductivity, (c) Thermal Conductivity, (d) The Figure of Merit

CONCLUSIONS

    In the present work, Ag-CoSb₃ nanocopmosites were successfully fabricated with
different Ag NP concentrations. Optimized SPS conditions were utilized to produce highly dense
pellets with preserved nanostructures. Though the increased ZT value is not as high as compared
to other class families of doped CoSb₃, ZT values are comparable to the undoped CoSb₃.[13]
Further investigation of higher concentration of Ag nanoparticles and hall measurements will be
performed to understand the dispersion behavior of Ag nanoparticles.

## ACKNOWLEDGMENTS

This work has been funded by EC-FP7 program under NEXTEC project and in part by the Swedish Foundation of Strategic Research - SSF.

## REFERENCES

1. D. M. Rowe, *CRC Handbook of Thermoelectrics*, (CRC, Boca Raton, FL, USA, 1995) p. 41-1.
2. C. Uher, *Skutterudites: Prospective Novel thermoelectrics Semiconductor and Semimetals* , T. M. Tritt ed., (Academic, San Diego, CA, 2001) p. 139.
3. J. S. Dyck, W. Chen, J. Yang, G. P. Meisner and C. Uher, *Phys. Rev. B: Condens. Matter.* **65**, 115204, (2002).
4. T. Caillat, A. Borshchevsky and J.-P. Fleurial, *J. Appl. Phys.* **80**, 4442, (1996).
5. G. S. Nolas, D. T. Morelli and T. M. Tritt, *Annu. Rev. Mater. Sci.* **29**, 89 (1999).
6. L. D. Chen, T. Kawahara, X. F. Tang, T. Goto, T. Hirai, J. S. Dyck, W. Chen, and C. Uher, *J. Appl. Phys.* **90**, 1864, (2001).
7. M. S. Toprak, C. Stiewe, D. Platzek, S. Williams, L. Bertini, E. Müller, C. Gatti, Y. Zhang, M. Rowe and M. Muhammed, *Adv. Funct. Mater.* **14**, 1189-1196, (2004).
8. M. Christensen, B. B. Iversen, L. Bertini, C. Gatti, M. S. Toprak, M. Muhammed, and E. Nishibori, *J. Appl. Phys.* **96**, 3148–3157, (2004).
9. C. Stiewe, L. Bertini, M. S. Toprak, M. Christensen, D. Platzek, S. Williams, C. Gatti, E. Müller, B. B. Iversen, M. Muhammed, and M. Rowe, *J. Appl. Phys.* **97**, 044317, (2005).
10. V. Y. Kodash, J. R. Groza, G. Aldica, M. S. Toprak, S. Li and M. Muhammed, *Scripta Materialia* **57**, 509–511, (2007).
11. A. Jacquot, M. Jaegle, H.-F. Pernau, J. König, K.Tarantik, K. Bartholomé, H. Böttner, Presented at *9th European Conference on Thermoelectrics, Thessaloniki,* **C-14-P,** (2011).
12. L.A. Stanciu, V.Y. Kodash, M. Crisan, M. Zaharescu and J.R. Groza, *J. Am. Ceram. Soc.* **84**, 983, (2001).
13. J. W. Sharp , E. C. Jones , R. K. Williams , P. M. Martin , and B. C. Sales, *J. Appl. Phys.* **78**, 1013 (1995).

Mater. Res. Soc. Symp. Proc. Vol. 1490 © 2012 Materials Research Society
DOI: 10.1557/opl.2012.1644

# Influence of Addition of Alumina Nanoparticles on Thermoelectric Properties of Higher Manganese Silicide

Takashi Itoh[1], Naoki Ono[1]
[1]Department of Materials, Physics and Energy Engineering, Nagoya University, Furo-cho, Chikusa-ku, Nagoya, 464-8603, Japan

## ABSTRACT

Higher manganese silicide (HMS) is a low-cost and eco-friendly thermoelectric material available for recovering waste heat of 500 to 900 K. In this research, we tried to uniformly disperse the alumina nanoparticles (ANPs) in the HMS matrix to reduce the thermal conductivity and to improve the thermoelectric performance. Influence of addition of ANPs on the thermoelectric properties was investigated. It was confirmed that ANPs were uniformly dispersed in the HMS grain boundary. The lattice thermal conductivity was reduced by adding ANPs. As a result, the maximum thermoelectric performance of $ZT$=0.58 was achieved at about 800 K by adding 1 vol% of ANPs. The performance of ANPs-added HMS was improved about 25 %.

## INTRODUCTION

Higher manganese silicide (HMS) is a p-type thermoelectric compound $MnSi_x$ with $x$ in the range from 1.67 to 1.75. Since both Mn and Si are nontoxic constituent elements and they exist in abundance as mineral resources, HMS can become a low-cost and eco-friendly thermoelectric material. It also attracts attention lately as a counterpart material of n-type $Mg_2Si$ compound that is similarly low-cost and eco-friendly. HMS is the generic name of $Mn_{11}Si_{19}$, $Mn_{26}Si_{45}$, $Mn_{15}Si_{26}$, $Mn_{27}Si_{47}$ and $Mn_4Si_7$ compounds. All of HMS compounds have a tetragonal crystal structure with unique very long unit cell along c-axis [1]. The manganese atoms construct a frame structure, and the silicon atoms construct a helical structure in the frame.

The HMS compound has been fabricated conventionally by an arc melting method, an induction heating method or several types of crystal growth methods. But, these fabrication methods require huge energy and/or long term, and they make an inhomogeneous microstructure. The HMS fabrication using powder metallurgical method can solve these disadvantages. For example, the combined method of mechanical alloying (MA) and pulse discharge sintering (PDS) has been used for fabricating HMS compound [2, 3]. As another powder metallurgical method, we proposed a new combined method of mechanical grinding (MG) and PDS [4, 5]. In this method, each of pure Mn and Si powders was mechanically ground using a ball milling equipment, and the HMS compound was synthesized from the ground powder mixture and simultaneously consolidated by PDS.

For improving the performance of thermoelectric materials without doping metal elements, the reduction in lattice thermal conductivity by phonon scattering is generally very effective. The grain refining and the uniform dispersion of nanoparticles cause the phonon scattering. Several trials on the nanoparticles addition to the thermoelectric bulk materials have been reported [6-11]. We have reported about the improvement of thermoelectric performance of $CoSb_3$ compound by uniformly dispersing nano-size materials (fullerenes, carbon nanotubes and alumina nanoparticles) [6-8]. Based on this knowledge, we attempted to apply the uniform dispersion of the alumina nanoparticles (ANPs) to the HMS, and investigated the influence of addition amount of ANPs on the thermoelectric properties.

## EXPERIMENT

As raw material powders, Mn powder (purity: > 99.9 %, particle size: < 75 μm) , Si powder (> 99.9 %, < 75 μm) made by Kojundo Chemical Lab. Co., Ltd., Japan and γ-ANPs (mean diameter: 50 nm) made by Praxair K.K., Japan were prepared. Each of Mn and Si powders was put into a alumina milling pot (capacity: 1000 ml) with alumina balls (diameter: 8 mm) with volume ratio of ball to powder of 10:1, and mechanically ground for 10 h in an argon atmosphere using a vibration ball milling (VBM) equipment (VS-1, Irie Shokai Co., Ltd., Japan). Then, the milled powders with particle size of 45 μm or less were individually collected by sieving in an argon atmosphere. The milled powders of Mn and Si with the starting composition $MnSi_{1.82}$ were weighed, and the γ-ANPs were added into the powders. For the ANPs-added samples, the powders were mechanically mixed using the planetary ball mill (P-6, Fritsch GmbH, Germany) at 150 rpm for 30 min in an argon atmosphere in order to uniformly disperse the ANPs. For the non-added sample (0 vol%ANPs), the milled Mn and Si powders were mixed using a rotary mixer (ANZ-51S, Nitto Kagaku Co., Ltd., Japan) at 100 rpm for 1 h in an argon atmosphere. The powder mixture was collected and dried within a glove box in an argon atmosphere. The powder samples mixed with 0, 0.5, 1 and 2 vol% of ANPs were prepared. The mixed powder sample was packed in a cylindrical graphite container of 20 mm in diameter. Then, the HMS compound samples were simultaneously synthesized and consolidated at 1173 K for 30 min in a vacuum under a pressure of 30 MPa using a PDS equipment (SPS-1050, Fuji Electronic Industrial Co., Ltd., Japan).

The sintered samples were cut into quadratic prism (3 x 3 x 13 mm), and the Seebeck coefficient and the electrical resistivity were measured using a thermoelectric measuring equipment (ZEM-2, Ulvac-Riko Inc., Japan). The disk-type specimens (φ10 x 1 mm) were prepared from the sintered samples, and the thermal conductivity was measured with a laser-flash equipment (TC-7000, Ulvac-Riko Inc., Japan). The sintered samples were characterized using an X-ray diffractometer (XRD) (RINT2500, Rigaku Co., Japan) with Cu Kα radiation and a scanning electron microscopy with energy dispersive x-ray spectroscopy (SEM-EDX) (SEMEDXIII Type-N, Hitachi High-Technologies Co., Japan).

## RESULTS AND DISCUSSION

### Phase Identification and Microstructure

Figure 1 shows the x-ray diffraction patterns of the synthesized samples with different addition amount of ANPs. In all samples, HMS phase was mainly identified. Thus, HMS compound was successfully synthesized by VBM-PDS process. A small amount of MnSi phase was also detected in the ANPs-added samples. It is believed that the MnSi phase appeared in the synthesis, because loss of Si was caused during mechanical mixing by planetary ball mill owing to easier adhesion of Si than Mn to the milling balls and pot.

SEM images with Al mapping of the ANPs-added samples are shown in Figure 2 (a)-(c). The points detecting Al element indicate the positions where ANPs exist. Though it is hard to recognize the positions of ANPs in Figure 2 (a)-(c), they are uniformly dispersed in the HMS matrix without agglomeration regardless of addition amount of ANPs. Figure 2(d) represents the high magnification SEM image with Al mapping of the 2.0 vol%ANPs-added sample. The grain boundaries of the HMS clearly appear by Al mapping. It was confirmed from this figure that the ANPs dispersed in the grain boundaries and the grain size was not reduced much by the mechanical mixing using the planetary ball mill.

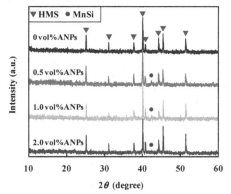

Figure 1. X-ray diffraction patterns of synthesized samples with different addition amount of ANPs.

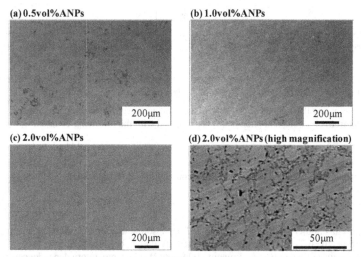

Figure 2. SEM images with Al mapping of ANPs-added samples; (a) 0.5 vol%, (b) 1.0 vol%, (c) 2.0 vol% and (d) 2.0 vol% (high magnification).

**Thermoelectric Properties**

Figure 3 (a) and (b) show the temperature dependences of the electrical resistivity and the Seebeck coefficient in the samples with different addition amount of ANPs, respectively. The electrical resistivity of all samples indicated similar tendency regardless of ANPs addition. The ANPs-added samples had the almost same resistivity regardless of addition amount of ANPs and its value was lower than that of the non-added sample (0 vol%ANPs) all over the temperature range. Though ANPs addition should generally increase the resistivity by the carrier scattering, the existence of MnSi phase having the lower resistivity than HMS would increase the carrier concentration and bring the reduction in resistivity. As the result, the addition amount of ANPs hardly affected the resistivity of HMS compound. There was a similar tendency between the resistivity and the Seebeck coefficient. Thus, appearance of MnSi phase by ANPs addition would increase the carrier concentration and lower the Seebeck coefficient.

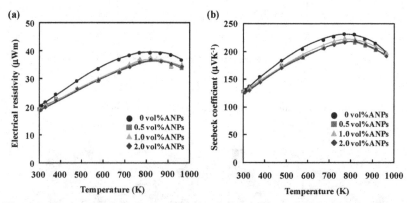

Figure 3. Temperature dependences of (a) electrical resistivity and (b) Seebeck coefficient in samples with different addition amount of ANPs.

Generally, the thermal conductivity $\kappa$ of thermoelectric semiconductors consists of contributions from electrons and phonons. It is expressed with the following equation:

$$\kappa = \kappa_{el} + \kappa_{ph}, \tag{1}$$

where $\kappa_{el}$ and $\kappa_{ph}$ are the electron and lattice thermal conductivities, respectively. According to the Wiedemann-Franz Law, the electron thermal conductivity $\kappa_{el}$ is calculated as follows:

$$\kappa_{el} = \frac{L_0 T}{\rho}, \tag{2}$$

where the Lorenz number $L_0$ is $2.45 \times 10^{-8}$ $V^2K^{-2}$, $T$ the absolute temperature and $\rho$ the electrical resistivity, respectively. The lattice thermal conductivity $\kappa_{ph}$ can be calculated from the measurement values of $\kappa$ and $\rho$ using Equations (1) and (2). The temperature dependences of the thermal conductivity and the lattice thermal conductivity in the samples with different addition amount of ANPs are shown in Figure 4(a) and (b), respectively. The thermal conductivity was lowered by ANPs addition. Though the thermal conductivity of alumina is about 30 $Wm^{-1}K^{-1}$ and much higher than HMS compound, the reduction would be mainly caused by phonon scattering at the ANPs in the grain boundaries. The addition amount of ANPs, however, hardly affected the thermal conductivity. The minimum conductivity of 1.72 $Wm^{-1}K^{-1}$ was obtained in the 0.5 vol%ANPs-added sample. It was 18% lower value than that of the non-added sample. The difference between the lattice thermal conductivities of the ANPs-added samples and the non-added sample was further widened because of reduction in electrical resistivity in the ANPs-added samples. The minimum value of the 0.5 vol%ANPs-added sample was 1.21 $Wm^{-1}K^{-1}$ and 26 % lower than that of the non-added one.

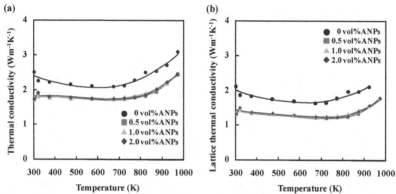

Figure 4. Temperature dependences of (a) thermal conductivity and (b) lattice thermal conductivity in samples with different addition amount of ANPs.

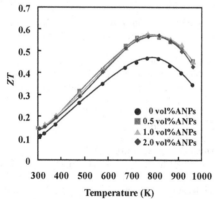

Figure 5. Temperature dependence of dimensionless figure of merit, $ZT$ in samples with different addition amount of ANPs.

The dimensionless figure of merit $ZT$ which represents the thermoelectric performance of the material is defined as follows;

$$ZT = \frac{S^2 T}{\rho \kappa},\qquad(3)$$

where $S$ is the Seebeck coefficient. The temperature dependences of $ZT$ in the samples with different addition amount of ANPs were estimated based on Equation (3) and they are shown in Figure 5. All samples have a similar tendency in the temperature dependence of $ZT$, that is, the maximum $ZT$ exists at around 800 K. The ANPs addition enhanced the thermoelectric performance of HMS because of the remarkable reduction in thermal conductivity. But the difference of addition amount of ANPs hardly affected the performance. The maximum $ZT$ of 0.58 was achieved in 1.0 vol%ANPs-added sample at 800 K, and the performance was improved about 25 % higher than that of the non-added sample.


CONCLUSIONS

We attempted to apply the uniform dispersion of the ANPs to the HMS, and investigated the influence of addition amount of ANPs on the thermoelectric properties. The HMS compound was successfully synthesized by VBM-PDS process, and the uniform dispersion of ANPs was achieved by mechanical mixing using the planetary ball mill. The existence of MnSi phase by ANPs addition resulted in the small reductions in both electrical resistivity and Seebeck coefficient. The phonon scattering caused by dispersing ANPs in the grain boundaries brought the large reduction in thermal conductivity. Though each thermoelectric property was hardly affected by addition amount of ANPs, the thermoelectric performance was enhanced by adding ANPs. The maximum $ZT$ of the 1.0 vol%ANPs-added sample was 0.58 at 800 K, and the performance was improved about 25 %.

REFERENCES

1.  R. De Riddera and S. Amelinckx, *Mat. Res. Bull.* **6**, 1223 (1971).
2.  M. Umemoto, Z. G. Liu, R. Omatsuzawa, K. Tsuchiya, *Mater. Sci. Forum* **343-346**, 918 (2000).
3.  T. Itoh, M. Yamada, *J. Electronic Materials* **38**, 925 (2009).
4.  M. Yoshikura, T. Itoh, *J. Jpn. Soc. Powder Powder Metallurgy* **57**, 242 (2010). (in Japanese)
5.  N. Ono, T. Itoh, *Proc. International Symposium on Materials Science Innovation for Sustainable Society*, Vol.2 pp.11-12, (2011).
6.  T. Itoh, K. Ishikawa, A. Okada, *J. Mater. Res.* **22**, 249 (2007).
7.  T. Itoh, M. Tachikawa, *Mater. Res. Soc. Symp. Proc.* Vol. 1314, Cambridge University Press, pp. Online mrsf10-1314-ll08-18 (6pp) (2011).
8.  T. Itoh, M. Matsuhara, *Mater. Trans.* **53**, 1801 (2012).
9.  X. Zhou, G. Wang, L. Zhang, H. Chi, X. Su, J. Sakamoto, C. Uher, *J. Mater. Chem.* **22**, 2958 (2012).
10. D. Cederkrantz, N. Farahi, K. Borup, B. Iversen, M. Nygren, A. Palmqvist, *J. Appl. Phys.* **111**, 023701 (2012).
11. Q. Zhang, H. Wang, Q. Zhang, W. Liu, B. Yu, H. Wang, D. Wang, G. Ni, G. Chen, Z. Ren, *Nano Lett.* **12**, 2324 (2012).

Mater. Res. Soc. Symp. Proc. Vol. 1490 © 2013 Materials Research Society
DOI: 10.1557/opl.2013.24

Diameter and Temperature Dependences of Phonon-Drag Magnetothermopower in
Bismuth Nanowires

Naomi Hirayama[1] , Akira Endo[2] and Naomichi Hatano[3]
[1]Tokyo University of Science, 2641 Yamazaki, Noda, Chiba 278-8510 Japan
[2]Institute for Solid State Physics, University of Tokyo, 5-1-5 Kashiwanoha, Kashiwa, Chiba
277-8581, Japan
[3]Institute of Industrial Science, University of Tokyo, 4-6-1 Komaba, Meguro, Tokyo 153-8505
Japan

ABSTRACT

We present theoretical calculations of the phonon-drag contribution to the Nernst
thermoelectric power $S_{yx}$ in Bismuth nanowires. We investigate the thermopower $S_{yx}$ with
diameters $L$ ranging from 22 to 900 nm at low temperatures (0.1 - 4.0 K) and high magnetic
fields (up to 16 T). We find that the peak of thermopower $S_{yx}$ around 14.75 T exhibits the size
effect in two different ways: for wires with $L \geq 200$ nm, the peak height increases with decreasing
$L$; for wires with $L < 200$ nm, on the other hand, the peak height rapidly decreases with decreasing
$L$. The dependence is accounted for by considering the contributions of discrete quantized
phonon modes. We also discuss the temperature dependence of $S_{yx}$.

INTRODUCTION

Low-dimensional thermoelectrical materials such as films, nanotubes, and nanowires have been
attracting considerable attention in both fundamental studies and industrial applications. The
efficiency of a thermoelectric material is evaluated by the dimensionless figure of merit $ZT$ with
$Z = S^2 \sigma / \kappa$, where $S$, $\sigma$, $\kappa$, and $T$ denote the longitudinal (Seebeck) thermoelectric power, electrical
and thermal conductivities, and the absolute temperature, respectively. The spatial confinement
in nano-structred materials is expected to suppress the excitation of phonons and the
accompanying thermal conductivity without affecting the electrical conductivity. The resulting
high figure of merit renders these materials promising candidates to be used in efficient and
environmentally benign thermoelectric energy-conversion devices. Despite the prospect,
however, none of such materials have flourished in practical success except for a small number
of specialized applications. In order to promote further development in thermoelectric devices,
fundamental understanding of the electric and thermal transport properties in nano-structured
materials is indispensable. It is not only of great theoretical interest but of practical importance in
facilitating the development of thermoelectrical energy generators. As an effort toward such
fundamental understandings, we here present theoretical calculations of the phonon-drag
contribution to the transverse thermomagnetic voltage, namely the Nernst voltage $S_{yx}$ in bismuth
nanowires and discuss the size and temperature dependences.

Bisumth has several interesting properties, such as the extraordinarily low carrier
densities ($\sim 10^{-5}$ per atom) and small effective masses ($\sim 0.06$ $m_0$ with $m_0$ the free electron mass).
The small carrier density is advantageous to the phonon-drag effect because carriers with small
Fermi momenta readily interact with phonons. In fact, the phonon-drag contribution is known to

play a dominant role in the Nernst voltage $S_{yx}$ of bismuth bulk under a quantizing magnetic field [2]. Moreover, bismuth is known as the material with the largest Nernst signal among semimetals and semiconductors around 4.2 K, where strong phonon drag is expected [3–5]. Owing to the small effective mass, the energy gap between Landau levels becomes large at a moderate magnetic field. In the phonon-carrier interaction at low temperatures, therefore, we need to consider only intra-Landau-level scatterings since the inter-Landau-level gap is much larger than the energy of available acoustic phonons. These characteristics let bismuth a material well-suited for investigating the phonon-drag Nernst effect.

In the present study, we calculate the phonon-drag Nernst voltage $S_{yx}$ of bismuth nanowires under a quantizing magnetic field applied along the trigonal axis of bismuth (z-axis). Hasegawa et al. [1] fabricated single-crystalline bismuth nanowires with diameter 80 - 500 nm and length over 1 mm, which is long enough to measure the thermoelectric voltage with high precision. Phonons in the transverse directions are quantized in such nanowires due to quantum confinement. The wire diameters are smaller than the mean free path of phonons at low temperatures but larger than the Fermi wave lengths of carriers (~20 - 30 nm). We therefore assume in the present study that the phonons travel one-dimensionally, while the carriers still behave three-dimensional bulk-like. We extend a model developed for bismuth bulk [2] to nanowires. We consider only holes as carriers, which have been shown to have the dominant contribution to $S_{yx}$ for the direction of the magnetic field considered in the present paper [2]. Our calculation has revealed that the peak of $S_{yx}$ exhibits non-trivial size effect, which is attributed to the quantization of the phonon modes.

## THEORY

We use a theory for a single-crystalline bismuth [2] that gives the phonon-drag Nernst thermopower $S_{yx}$. In the theory, the thermopower $S_{yx}$ is expressed as the integration with respect to the wave vectors of phonons $\mathbf{q} = (q_x, q_y, q_z)$ as follows:

$$S_{yx} = -\frac{1}{(2\pi)^4}\frac{e\rho_{xx}L_x}{2k_BT^2\rho\hbar}D^2m_z^*\sum_\sigma\sum_{n,n'}\int d\mathbf{q}\frac{q_x^2q}{q_z}N_q^{(0)}I_{n,n'}(q)f[E(n,k_{z0},\sigma)]\{1-f[E(n,k_{z0},\sigma)]\}, \quad (1)$$

$$E(n,k_{z0},\sigma) = \hbar\omega_h\left(n+\frac{1}{2}\right)+\frac{1}{2}\sigma g\mu_B B+\frac{\hbar^2k_{z0}^2}{2m_z}, \quad (2)$$

$$k_{z0} = \frac{m_z}{\hbar q_z}\left[\omega_h(n-n')+\omega_q\right]-\frac{1}{2}q_z, \quad (3)$$

$$I_{n,n'} = \left|\int_{-\infty}^{\infty}\phi(x-X_0;n')e^{-i\mathbf{q}\cdot\mathbf{r}}\phi(x;n)\right|^2. \quad (4)$$

with a magnetic field $B$ applied in the $z$ (trigonal)-direction, the effective mass $m_z = 0.6667m_0$ ($m_0$ denotes the free electron mass), the density $\rho = 9.75\times10^3$ kg/m$^3$, the diagonal resistivity $\rho_{xx}$, the system size in the x-direction $L_x$, the deformation potential $D = 1.2$ eV [7], the equilibrium Bose distribution of phonons $N_q$, the Boltzmann constant $k_B$, the Landau indices $n$ and $n'$, and the spin index $\sigma = \pm 1$. Here, the frequency of acoustic phonons and the cyclotron frequency of holes are given as $\omega_q = v_s q$ and $\omega_h = eB/m_x$, respectively, where the group velocity of phonons $v_s = 2.0 \times 10^3$ m/s and $m_x = 0.06289 m_0$. We consider only the intra-Landau-level scattering ($n = n'$) in the phonon-carrier interactions, as mentioned earlier. In such cases, the quantity $I_{n,n'}$ can be written in the following form;

$$\left| I_{n,n'}(A_q) \right|^2 = \left[ L_n \left( \frac{A_q}{2} \right) \right]^2 \exp\left( -\frac{A_q}{2} \right), \tag{5}$$

$$A_q = \left( \frac{\chi_h}{l_h} \right)^2 + (q_x l_h)^2 = \frac{\hbar}{q_e B}(q_x^2 + q_y^2) = \frac{\hbar}{q_e B} q^2 \sin^2 \theta, \tag{6}$$

where $L_n(x)$ is the Laguerre polynomial.

In the present study, we consider a bismuth nanowire with the length $L_x = 2.2$ mm [2, 6] oriented along the $x$-direction (perpendicular to the trigonal axis), having a square cross section with sides $L_y = L_z = L$. The wave-vector components in the $y$- and $z$-directions are given by $q_y = n_y \pi/L$ and $q_z = n_z \pi/L$ ($n_y, n_z = 0, 1, 2,...$), respectively, since the phonons are quantized owing to the confinement into the wire geometry. We convert the equation for a bulk system to a nanowire with quantized phonons by substituting the summation for the integration with respect to q in Eq. (1) in the $y$- and $z$-directions, i.e., $\int d\mathbf{q} \rightarrow (\pi^2/d^2) \sum_{n_y, n_z = 1, 2, \cdots} \int dq_x$. Equation (1) is thus rewritten as follows:

$$S_{yx} = \left( \frac{\pi}{L} \right)^2 \sum_{n_y = 1, 2, \cdots} \sum_{n_z = 1, 2, \cdots} U(q_y, q_z), \tag{7}$$

$$U(q_y, q_z) = -\frac{1}{(2\pi)^4} \frac{e \rho_{xx} L}{2 k_B T^2 \rho \hbar} D^2 m_z^*$$

$$\times \sum_{\sigma} \sum_{n = n'} \int dq_x \frac{q_x^2 q}{q_z^2} N_q^{(0)} I_{n,n}(q) f\left[ E(n, k_{z0}, \sigma) \right] \left\{ 1 - f\left[ E(n', k_{z0}, \sigma) \right] \right\}. \tag{8}$$

The function $U(q_y = n_y \pi/L, q_z = n_z \pi/L)$ is the contribution of the discrete phonon mode with $n_y$ and $n_z$ to the Nernst voltage $S_{yx}$.

## DISCUSSION

Figures 1(a) and 1(b) show the Nernst thermopower $S_{yx}$ of the nanowires with the thickness $L$ ranging from 22 to 900 nm, at 0.28 K. We find that the peak of the thermopower $S_{yx}$ around 14.75 T (where the Landau level of the holes with $n = 1$ and $\sigma = 1$ crosses the Fermi energy) exhibits the following size dependence; For 200 nm $\leq L \leq$ 900 nm (Fig. 1(a)), the peak becomes higher and narrower with decreasing $L$. The maximum peak height is achieved at $L = 200$ nm, where the peak height is roughly four times larger than that in the bulk. For smaller diameters 22 nm ≤ $L$ ≤ 90 nm (Fig. 1(b)), on the other hand, the peak shrinks with decreasing $L$.

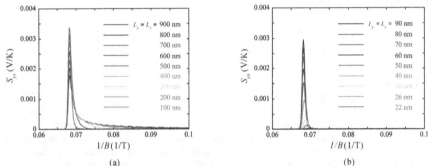

(a)                                                    (b)

**Figure 1.** Nernst voltage $S_{yx}$ with wire thickness **(a)** $L_y = L_z =$ 100 nm - 900 nm and **(b)** $L_y = L_z =$ 10 nm - 90 nm.

The contribution of phonons with $q_y$ and $q_z$ to the Nernst voltage, $U(q_y, q_z)$, has peaks on the $q_y - q_z$ plane, as shown in Figure 2. The total Nernst voltage $S_{yx}$ is obtained by summing up $U(q_y, q_z)$ with $q_y = n_y\pi/L$ and $q_z = n_z\pi/L$ over the quantized phonon modes indicated by blue dots in Fig. 2(b). The change in $L$ alters the positions of the phonon modes. The voltage $S_{yx}$ becomes large when phonon modes lie close to the peaks in $U(q_y, q_z)$.

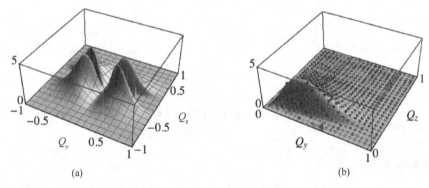

(a)                                                    (b)

**Figure 2. (a)** Contribution $U(q_y, q_z)$ of phonons with $(q_y, q_z)$ to the Nernst voltage $S_{yx}$. Dimensionless wave numbers of the phonons, $Q_y = q_y l$ and $Q_y = q_y l$, are used in the plot with $l = \sqrt{\hbar/eB}$ the magnetic length. **(b)** The part $Q_y > 0$, $Q_z > 0$ in Fig. 2(a) superposed with the discretized phonon states for $L = 500$ nm.

The mechanism of the size dependence of $S_{yx}$ is schematically illustrated in Fig. 3. When the phonon mode near the origin ($n_y = n_z = 1$) moves passing through the peak of $U(q_y, q_z)$ with decreasing $L$, the voltage $S_{yx}$ rises while approaching the peak of $U(q_y, q_z)$ (Fig. 3(a)), reaches the maximum when the mode hits the peak (Fig. 3(b)), and then decreases with the further decrease in $L$ (Fig. 3(c)). This explains the size-dependent behavior mentioned above.

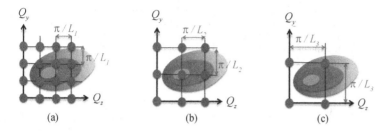

(a)          (b)          (c)

**Figure 3. (a)-(c)** Schematic diagram explaining the increase and the decrease of the peak height with decreasing $L$, $L_1 > L_2 > L_3$.

Figures 4(a) and 4(b) show the temperature dependence of $S_{yx}$ for the sample with $L = 500$ nm. The Nernst thermopower $S_{yx}$ monotonically decreases with decreasing $T$. This drop of $S_{yx}$ is caused by the suppression of phonon drag due to the reduction of the number of phonons at low temperatures. Similar temperature dependence has been also reported for a bismuth bulk in experimental [6] and theoretical [2] studies.

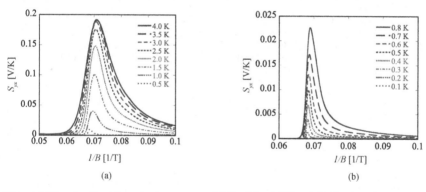

(a)          (b)

**Figure 4.** Nernst voltage $S_{yx}$ at the temperatures **(a)** $T = 0.5 - 4.0$ K and **(b)** $T = 0.1 - 0.8$ K.

## CONCLUSIONS

We have investigated the phonon-drag Nernst electromotive force $S_{yx}$ of Bismuth nanowires assuming that the acoustic phonons are quantized owing to the confinement into the wire geometry, such that the wave-vector components in the $y$- and $z$-directions are given by $q_y = n_y \pi / L$ and $q_z = n_z \pi / L$ ($n_y, n_z = 0, 1, 2, \ldots$). We found that the peak of the thermopower $S_{yx}$ around 14.75 T (where a Landau level of the holes with $n = 1$ and $\sigma = +1$ crosses the Fermi energy) exhibits the non-trivial size effect; the peak becomes higher with decreasing $L$ down to ~200 nm; for smaller $L$, on the other hand, the peak shrinks with decreasing $L$. The increase and decrease of the peak height can be explained by the passing of the phonon mode $(\pi/L, \pi/L)$ through the peak in the integrand $U(q_y, q_z)$ with decreasing wire thickness $L$.

## ACKNOWLEDGMENTS

The authors are grateful to Prof. H. Nakamura and Prof. Y. Hasegawa for valuable discussions.

## REFERENCES

1.  Y. Hasegawa, D. Nakamura, M. Murata, H. Yamamoto, T. Komine, T. Taguchi, and S. Nakamura, J. Elec. Mat. **40**, 1005 (2011).
2.  M. Matsuo, A. Endo, N. Hatano, H. Nakamura, R. Shirasaki, and K. Sugihara, Phys. Rev. B **80**, 075313 (2009).
3.  K. Behnia, J. Phys.: Condens. Matter **21**, 113101 (2009).
4.  K. Behnia, M.-A. M_easson, and Y. Kopelevich, Phys. Rev. Lett. **98**, 076603 (2007).
5.  D.M. Jacobson, J. Appl. Phys. **45**, 4801 (1974).
6.  K. Behnia, M.-A. M_easson, and Y. Kopelevich, Phys. Rev. Lett. **98**, 166602 (2007).
7.  K. Walther, Phys. Rev. **174**, 782 (1968).

Mater. Res. Soc. Symp. Proc. Vol. 1490 © 2013 Materials Research Society
DOI: 10.1557/opl.2013.317

# Thermoelectric Measurements of Ni Nanojunctions

See Kei Lee[1], Ryo Yamada[1] and Hirokazu Tada[1]
[1]Graduate School of Engineering Science, Osaka University, Toyonaka, Osaka, Japan.

## ABSTRACT

We investigated the thermoelectric voltage (TEV) of atomic contacts of nickel (Ni) by using a scanning tunneling microscope. The TEV of nanoscale junctions show fluctuation in stepwise manner. Histogram analysis of TEV observed in the Ni point contact with the conductance of 1.2 $G_0$ ($G_0 = 2e^2/h$ is the quantum of charge conductance) revealed multiple voltage peaks at larger and smaller values observed at conductance of 2.5 $G_0$, which showed a single sharp voltage peak. Fluctuation observed in our results suggest that there is transition of the transport channel distribution caused by the thermal motion of Ni atoms.

## INTRODUCTION

Studies on thermoelectric effects (TE) in atomic and molecular junctions have attracted much attention because these junctions can realize a narrow distribution of the energy of the electrons participating in the transport process which can contribute to maximum thermoelectric efficiency [1-2]. The efficiency of thermoelectric devices is represented as:

$$ZT = \frac{\sigma S^2 T}{\left( \kappa_e + \kappa_l \right)}$$

where $\sigma$ is the electrical conductivity, $S$ is the Seebeck coefficient, and $\kappa_e$ and $\kappa_l$ are the electronic and lattice part of the thermal conductivity, respectively. A greater $ZT$ would indicates a greater efficiency of the device.

Nanoscale materials have shown much promise as good thermoelectric material because of two reasons [3-6]: i) Low-dimensionality and quantum size effects could improve the Seebeck coefficient and ii) Small feature sizes enhance phonon scattering on nanoscale interfaces and reduce thermal conductivity. Application of nanoscale thermoelectric devices can be beneficial as it can be integrated into chip sets and converting the accumulated waste heat into usable electric energy [7].

Electronic transport phenomena in low-dimensional systems can be roughly divided into two categories: ballistic transport and diffusive transport. This two transport are realized when the wire length is shorter and longer, respectively, than the mean free path of electron. It had been discussed that a higher thermoelectric effiency value can be expected for a system showing ballistic transport [6]. Seebeck coefficient of the quantized transport system is predicted to change oscillatory as a function of the conductance [7], showing the maximum Seebeck coefficient at $(n+1/2)G_0$. This behavior is experimentally observed in 2D electron gas system [8].

The thermoelectic characteristics of atomic junction is interesting because the junction shows ballistic transport even at room temperature. The thermorelectric behaviors of gold, silver and Cu atomic contacts were investigated by a mechanically controlled break junction [9]. The Seebeck coefficient randomly scattered around 0 at $G = nG_0$ and small positive peaks were observed around $G = (n+1/2)G_0$ (n; integer), which resembled with what expected by the theory.

In this work, we observed the thermoelectric voltage of Ni point contacts.Because the charge transport channels are not fully opened and mixture of multiple valence channels [10-12]. Our result suggests that this complex nature of the Ni junction could result in the thermoelectric behaviors which cannot be explained by simple ballistic model.

**EXPERIMENT**

The schematic figure of our experimental setup is shown in Figure 1. The experiment was conducted at room temperature in an argon gas (Ar) by a home-build scanning tunneling microscope. The difference in temperature between a tip and a substrate, $\Delta T$, was created by controlling the temperature of the substrate by using a Peltier device. The tip was mounted on a copper (Cu) block and kept around room temperature. The temperatures of the Cu blocks holding tip and substrate were monitored by Si diode sensors. A Ni tip was prepared by cutting the Ni wire with a diameter of 0.25 mm. Ni substrate was prepared by thermal evaporation on Mica. The Ni tip was brought close to the Ni thin film substrate while applying a bias voltage of 50 mV until the threshold current value of 2.5 $G_0$ and 1.2 $G_0$ was reached [13, 14]. Then, the bias voltage and current amplifier were disconnected, and the voltage amplifier was connected instead (as shown Fig. 1) to measure the tip-substrate TEV induced by the $\Delta T$. After the voltage measurement, the electrical conductance of the junction was measured again without feedback to confirm the stability of the junction. If the junction was reconfirmed to exist, the TEV was measured again. We usually repeated these cycles for 3 times. The voltage of the contact was determined from the voltage histogram created from the data obtained.

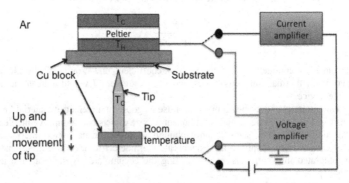

**Figure 1.** Schematic of the voltage measurement setup.

## RESULTS AND DISCUSSION

Fig. 2(a) shows the current and voltage, which were alternatively measured, as a function of time when the tip position was held at 2.5 $G_0$. It was shown that the current went back to the initial value after the voltage measurement and the voltage value was constant. However, when the tip was held at G = 1.2 $G_0$ (Fig. 2(b)), stepwise fluctuation was observed.

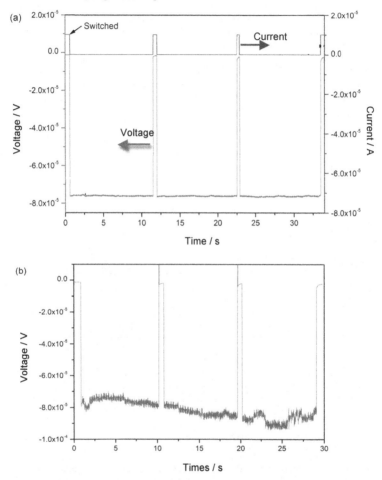

**Figure 2.** Voltage (and current) as a function of time graph when the tip was held at (a) 2.5 $G_0$ and (b) 1.2 $G_0$ at $\Delta T$ = 10.5 K.

Figure 3 shows voltage histograms obtained at different $\Delta T$ when the tip was held at $G =$ 1.2 $G_0$ and 2.5 $G_0$. The histograms taken for the junction of $G = 2.5$ $G_0$ showed single sharp voltage peak at all temperature differences. The Seebeck coeficcient calculated from the data is - 10.14 $\mu$V/K. The histogram taken for the junction of $G = 1.2$ $G_0$ showed broadened and multiple peaks, which were originated to the fluctuation of the voltage. The fluctuation of the Seebeck coefficient was reported in experiment [9] and theory [15]. However, the Seebeck coefficient fluctuated around 0 V in previous repots whereas our result shows fluctuation around multiple center values. This discrepancy could be attributed to the different mechanism of fluctuation. In previous repots, the fluctuation of the Seebeck coefficient is attributed to the scattering phenomena at the junction. In case of Ni contact, the Seebeck coefficient can be related to the electronic structure and transmission function of the junction. Ienaga $et$ $al$ measured the differential conductance of the Ni junctions and it has negative slope around V = 0 above 200 K [16]. Since the Seebeck coefficient is proportional to the $d\ln G/dV$, the negative slope of the differential conductance around V = 0 gives negative Seebeck coefficient, which qualitatively agree with the present result.

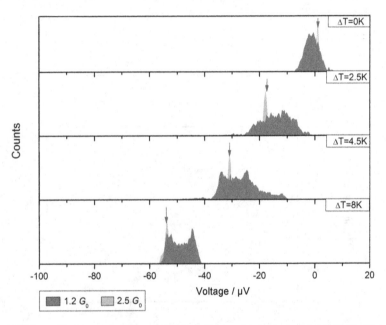

**Figure 3.** Voltage histograms at different $\Delta T$ when tip was fixed at setpoint 1.2 $G_0$ (dark) and 2.5 $G_0$ (light). Red arrows indicate the peak position observed when the setpoint = 2.5 $G_0$.

It is reported that the Ni contact with different diameters and atomic arrangements can show the conductance around 1 $G_0$. Therefore, the fluctuation and multiple peak of the observed thermovoltage can be attributed to the change of the transport channel due to the change of the atomic configuration caused by thermal fluctuation although the conductance was not significantly changed. [17].

## CONCLUSIONS

Ni point contact with the conductance of 1.2 $G_0$ shows multiple thermoelectric voltages around the value observed of the Ni contact with the conductance of 2.5 $G_0$, which shows single sharp peak. The fluctuation could be related to the transition of the transport channel distribution caused by thermal motion of atoms. Our results shows the usefulness of thermoelectric voltage on the analysis of transport mechanism of the point contact and enhancement of the thermoelectric power is possible by tuning the atomic arrangement of the point contact.

## ACKNOWLEDGMENTS

This work was supported by a Grant-in-Aid for Scientific Research on Innovative Areas, "Emergence in Chemistry" (21111514) from the Ministry of Education, Culture, Sports, Science, and Technology of Japan.

## REFERENCES

1. J. A. Malen, S. K. Yee, A. Majumdar and R. A. Segalman, *Chemical Physics Letters* **491**, 109-122 (2010).
2. G. D. Mahan and J. O. Sofo, *Proc. Natl Acad. Sci. USA* **93** (15) 7436. (1996).
3. T. C. Harman, P. J. Taylor, M. P. Walsh, and B. E. LaForge, *Science* **297** (5590), 2229 (2002).
4. L. D. Hicks and M. S. Dresselhaus, *Phys. Rev. B* **47** (24) 16631, 1993.
5. R. Venkatasubramanian, E. Siivola, T. Colpitts and B. O' Quinn, *Nature* **413**, 597-602, 2001
6. N. Neophytou and H. Kosina, *Journal of Computational Electronics* **11**: 29-44, (2012).
7. P. Streda, *J. Phys.: Condens. Matter* **1**, 102551027 (1989)
8. H. van Houtent, L. W. Molenkampt, C. W. J. Beenakkert and C. T. Foxon, *Semicond. Sci. Technol.* **7**, 8215-8221 (1992)
9. B. Ludoph and J. M. van Ruitenbeek, *Phys.Rev.B* **59**, 12290 (1999).
10. M. R. Sullivan, D. A. Boehm, D. A. Ateya, S. A. Hua and H. D. Chopra, *Physical Rev. B* **71**, 024412 (2005
11. M. R. Calvo, J. Ferna´ndez-Rossier, J. J. Palacios, D. Jacob, D. Natelson and C. Untiedt, *Nature Lett.* **458**, 1150-1154 (2009)
12. F. Pauly, M. DreherJ, M. Häfner, J. C. Cuevas, P. Nielaba, *Phys. Rev. B* **74**, 235106(2006).
13. P. Reddy, SY Jang, R. A. Segalman and A. Majumdar, *Science* **315**, 1568 (2007).
14. S. K. Yee, J. A. Malen, A. Majumdar and R. A. Segalman, *Nano Lett.* **11**, 4089 (2011).
15. F. Pauly, J. K. Viljas, M. Burkle, M. Dreher, P. Nielaba, and J. C. Cuevas, *Phys. Rev. B* **84**, 195420 (2011). – molecular dynamics thermopower and conductance calculation
16. K. Ienaga, N. Nakashima, Y. Inagaki, H. Tsujii, S. Honda, T. Kimura, and T. Kawae, *Phys.*

*Rev. B* **86**, 064404 (2012).

17. J. A. Malen, P. Doak, K. Baheti, T. D. Tilley, A. Majumdar and R. A. Segalman, *Nano Lett.* **9** (10) 3406 (2009).

Mater. Res. Soc. Symp. Proc. Vol. 1490 © 2012 Materials Research Society
DOI: 10.1557/opl.2012.1645

# Tantalum Nitride for Copper Diffusion Blocking on Thin Film (BiSb)2Te3

H. H. Hsu[1], C. H. Cheng[*,2], C. K. Lin[1], K. Y. Chen[1] and Y. L. Lin[1]
[1] Green Energy and Environment Research Lab., Industrial Technology Research Institute, Hsinchu, Taiwan, R.O.C.
[2] Department of Mechatronic Technology, National Taiwan Normal University, Taipei, Taiwan, R.O.C

## ABSTRACT

This study demonstrates the feasibility of introducing a TaN thin film as a copper diffusion barrier for $p$-type (BiSb)$_2$Te$_3$ thermoelectric material. Compared to conventional Ni diffusion barrier, remarkably little void generation in Cu bulk or near Cu/TaN interface originated from Cu penetration is observed for TaN barrier after suffering the thermal budget of close to soldering. Diffusion behaviors of the barriers were analyzed by transmission electron microscopy (TEM) and energy dispersive spectrometry (EDS) to make a deep understanding in clarifying interface diffusion effects among the Cu electrode, the barrier layer, and the (BiSb)$_2$Te$_3$ thermoelectric layer.

## INTRODUCTION

The solid-state thermoelectric devices are exploited for the direct conversion between heat and electric energy due to their advantages of no moving parts, no refrigerant needed, and high reliability [1]. Among thermoelectric materials, bismuth telluride compounds have been attracting considerable attention on micro-cooler applications because of superior thermoelectric characteristics [2,3]. However, copper (Cu) diffusion initiated by soldering process would degrade the thermoelectric properties due to the formation of copper telluride [4] or changing carrier concentration of thermoelectric materials [5,6]. Although quite a few studies have been reported to solve the diffusion issues for bulk thermoelectric materials [7-9], little research including observation of diffusion behavior was carried out for micro-thermoelectric devices. Many anti-diffusion materials such as Au, Ag, Ni, Ta, TiN and TiW were reported in a previous study [10], but these turned out not working for $p$-type Bi$_{0.5}$Sb$_{1.5}$Te$_3$ thin films. It was demonstrated that no thermoelectric cooling functionality might be due to diffusion of titanium into the thermoelectric elements [11]. To improve these issues, we proposed a robust metal nitride, namely tantalum nitride, to replace conventional Ni and investigated their stabilities against Cu diffusion.

## EXPERIMENT

The experimental procedure is concisely described as follows. The 170-nm-thick $p$-type (BiSb)$_2$Te$_3$ films were deposited by a multi-chamber sputter system with a processing temperature of 150°C on 500-nm-thick SiO$_2$/Si substrate. After film deposition, in-situ annealing at 250°C was performed to enhance film quality. Subsequently, a lift-off process was used to define transfer length method (TLM) pattern and then followed by metal depositions of 110-nm-thick Ni or TaN as Cu diffusion barrier layers. After that, 300-nm-thick Cu was deposited by sputtering as top electrodes. Finally, the post metal annealing at 200°C close to soldering process in thermoelectric module was applied for all devices to observe diffusion phenomena among layers. These samples were inspected by energy dispersive spectrometry (EDS), transmission electron microscopy (TEM) and grazing incidence X-ray diffraction (GIXRD) to clarify interface diffusion effects among Cu, barrier layers, and $(BiSb)_2Te_3$ thermoelectric layer. Current-voltage ($I$-$V$) curve measurement was performed by HP4156C semiconductor parameter analyzer.

## DISCUSSION

The two terminal multiple contact resistor method, also known as the TLM, was used to measure the contact resistance ($R_c$) and to extract the contact area-independent specific contact resistivity ($\rho_c$). Figure 1(a) and 1(b) show the schematic plot of the diffusion barrier layer on $(BiSb)_2Te_3$ layer and practical optical micrograph (OM) image of TLM pattern with several metal pads separated by 10, 20, 30, 40 μm after a lift-off process, respectively. From TLM measured results, TaN metal film has larger specific contact resistivity (3394 μΩcm$^2$) than Ni (170 μΩcm$^2$) which may be due to nitrogen incorporation or ion bombardment damage of surface during sputtering. However, its robust thermal stability and high IC integration capability have been demonstrated in back-end interconnection process with a thermal budget of 400°C. Furthermore, reduced specific contact resistivity by 2 to 4 orders of magnitude at barriers/thermoelectric layers can be achieved using solvent-cleaned or plasma treatment prior metal deposition and post metal deposition annealing [12]. More importantly, the measured $I$-$V$ characteristics in Figure 2(a) give evidence of Ohmic conduction, favored for the interface between [Ni or TaN] barriers and $(BiSb)_2Te_3$ layers. Figure 2 (b) presents the total resistance as a function of metal pad spacing for TLM measurement.

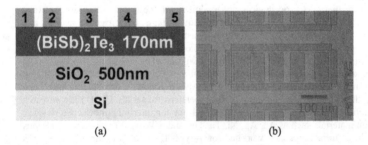

(a)                          (b)

**Figure 1.** (a) Cross-sectional schematic and (b) top view of TLM pattern on $(BiSb)_2Te_3$ layer for the specific contact resistivity measurement.

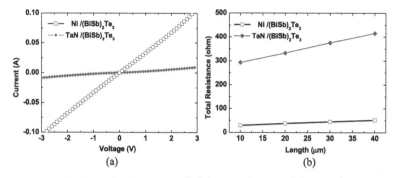

(a)                                    (b)

**Figure 2.** (a) Measured I-V characteristics on TLM and (b) total resistance as a function of TLM pad spacing for Ni and TaN contacts on $(BiSb)_2Te_3$ layer.

For a more detailed investigation of Cu barrier functionality, EDX (Figure 3) was performed to analyze the $Ni/(BiSb)_2Te_3$ interfacial region of the Ni barrier sample with 200°C post metal annealing. According to the EDX results, a remarkable Cu signal detected among the thermoelectric layer indicates Ni barrier layer cannot effectively prevent copper diffusion into $(BiSb)_2Te_3$ layer. In Figure 4, the formation of $NiTe_2$ compound is confirmed by the GIXRD inspection due to large Gibbs free energy, which can be supported by thermodynamic data [13]. These experimental results reveal that the severe Cu interdiffusion and reaction with $(BiSb)_2Te_3$ material could largely change material properties and thereby affect thermoelectric characteristics to result in the long-term reliability issue.

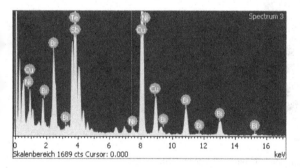

**Figure 3.** EDX spectra at the location of $Ni/(BiSb)_2Te_3$ interfacial region.

147

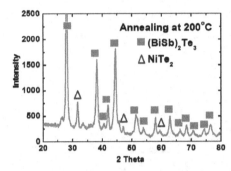

**Figure 4.** GIXRD spectra for Ni/(BiSb)$_2$Te$_3$ sample, showing NiTe$_2$ compound formation at 200°C.

Even after 200°C annealing treatment, TaN barrier layer demonstrates good step coverage on the rough (BiSb)$_2$Te$_3$ surface, compared to Ni barrier. Less voids or interfacial layer formation can be found from TEM image of Figure 5. To further evaluate the stability of TaN barrier layer on (BiSb)$_2$Te$_3$, the GIXRD analysis using Cu K$_\alpha$ radiation at 0.5° incident angle was generally performed to analyze phase transformation with temperature and also could be applied to detect the de-bonding or re-crystallized metal atoms, responsible for interface diffusion phenomenon. As seen in GIXRD data, no apparent telluride-based metal compound formation is found in robust TaN barrier case, implying the thermodynamic stability of TaN barrier is better than that of Ni, which may point to potential for application in thermoelectric modules.

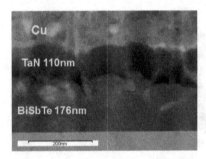

**Figure 5.** Cross-sectional TEM image of TaN barrier layer on (BiSb)$_2$Te$_3$.

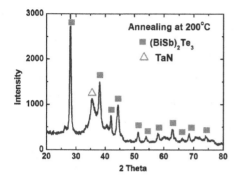

**Figure 6.** GIXRD spectra for TaN/(BiSb)$_2$Te$_3$ sample.

## CONCLUSIONS

Compared to commercial (BiSb)$_2$Te$_3$ technology, using a Ni diffusion barrier, the TaN barrier showed little void generation during an anneal at the soldering temperature. The improvement on copper diffusion blocking could be attributed to the nature of robust metal nitride with very strong metal-nitrogen bonding that was demonstrated in production lines of IC back-end interconnection or DRAM process with high-temperature thermal budget of 400°C. Subjects for future study, including the adhesion issue of Cu/TaN/(BiSb)$_2$Te$_3$ multi-layer, the minimum barrier thickness to optimize the thermoelectric performance, reduction of electrical contact resistance, and the thermodynamic behavior between metal nitride and thermoelectric layer are worth investigating.

## ACKNOWLEDGMENTS

The financial support provided by Bureau of Energy, Ministry of Economic Affairs, R.O.C. is gratefully acknowledged.

## REFERENCES

1. F. J. DiSalvo, Science, **285**, 703 (1999).
2. A. Majumdar, Science, **303**, 777□(2004).
3. J. P. Fleurial, A. Borshchevsky, M.A. Ryan, W. Phillips, E. Kolawa, T. Kacisch and R. Ewell, 16th International Conference on Thermoelectrics, IEEE (1997)
4. T Kacsich, E Kolawa, J P Fleurial, T Caillat and M-A Nicolet, J. Phys. D: Appl. Phys., **31**, 2406 (1998).
5. Y. C. Lan, D. Z. Wang, G. Chen and Z. F. Ren, Appl. Phys. Lett. **92**, 101910 (2008).
6. J. Bludska´, S. Karamazov, J. Navra´til, I. Jakubec, J. Hora´k, Solid State Ionics, **171**, 251 (2004).
7. W. P. Lin, D. E. Wesolowski, C. C. Lee, J Mater Sci: Mater Electron, **22**, 1313 (2011).
8. T. Y. Lin, C. N. Liao and A. T. Wu, Journal of Electronic Materials, **41**, 153 (2012)

9. R. P. Gupta, O. D. Iyore, K. Xiong, J. B. White, K. Cho, H. N. Alshareef and B. E. Gnade, Electrochemical and Solid-State Letters, **12**, H395 (2009).
10. N. H. Bae, S. Han, K. E. Lee, B. Kim and S.T. Kim, Current Applied Physics, **11**, S40 (2011).
11. L. W. da Silva and M. Kaviany, Journal of Microelectromechanical Systems, **14**, 1110 (2005).
12. R. P. Gupta, K. Xiong, J. B. White, K. Cho, H. N. Alshareef and B. E. Gnade, J. Electrochem. Soc., **157**, H666 (2010).
13. A. D. Mah and L. B. Pankratz, US Bureau of Mines, Bulletin, **668**, 125 (1976)

**Thermoelectric Properties and Applications**

Mater. Res. Soc. Symp. Proc. Vol. 1490 © 2013 Materials Research Society
DOI: 10.1557/opl.2012.1730

# Practical electrical contact resistance measurement method for bulk thermoelectric devices

Rahul P. Gupta, Robin McCarty, Jim Bierschenk, and Jeff Sharp
Marlow Industries Inc., a subsidiary of II-VI Inc., Dallas TX 75238

## ABSTRACT

As thermoelectric (TE) element length decreases, the impact of contact resistance on TE device performance grows more significant. In fact, for a TE device containing 100-$\mu$m tall $Bi_2Te_3$ TE elements, the figure of merit ratio ($ZT_{Device}/ZT_{Material}$) drops from 0.9 to 0.5 as the contact resistivity increases from 5 x $10^{-07}$ to 5 x $10^{-06}$ $\Omega$-cm$^2$. To understand the effects of contact resistance on bulk TE device performance, a reliable experimental measurement method is needed. There are many popular methods to extract contact resistance such as Transmission Line Measurements (TLM) and Kelvin Cross Bridge Resistor method (KCBR), but they are only well-suited for measuring metal contacts on thin films and do not necessarily translate to measuring contact resistance on bulk thermoelectric materials. The authors present a new measurement technique that precisely measures contact resistance (on the order of 5 x $10^{-07}$ $\Omega$-cm$^2$) on bulk thermoelectric materials by processing stacks of bulk, metal-coated TE wafers using TE industry standard processes. One advantage of this technique is that it exploits realistic TE device manufacturing techniques and results in an almost device-like structure, therefore representing a realistic value for electrical contact resistance in a bulk TE device. Contact resistance measurements for metal contacts to n- and p-type $Bi_2Te_3$ alloys are presented and an estimate of the accuracy of the measurements is discussed.

## INTRODUCTION

Cooling power density quantifies the capacity of a TE device to pump heat. High watt density devices can dramatically change the historical standards of cost per watt of heat pumped. Maximum cooling power density per unit area for a single element ($Q_{max}$) is derived by differentiating total heat flux (Q) for Peltier cooling [1] and is given by Equation 1,

$$Q_{max} = \frac{1}{2L}\left[\frac{\alpha^2 T_c^2}{\left(2\rho + \frac{4\rho_{cont}}{L}\right)} - k\Delta T\right] \qquad (1)$$

where L is the device leg length, $\rho_{cont}$ is the contact resistance, $\rho$ is the bulk resistivity, k is the thermal conductivity, $T_c$ is the cold side temperature, and $\alpha$ is the Seebeck coefficient. From Equation 1, cooling power density is inversely proportional to the leg length, L of the device; therefore cooling power density can be increased by reducing the leg length. Bulk material is currently the predominant commercial option to obtain the required performance because of the high cost of thin film material processing and limitations on film thickness. Micro TE coolers, bulk TE coolers with very small element lengths are becoming the norm in the telecommunications industry where the TE cooler temperature stabilizes the laser diodes that send the optical signals down the fiber optic cables [2]. The impact of contact resistance on micro TE cooler performance is quite significant for devices with element length of 200 um or less.

One of the requirements to quantify the effects of contact resistance on a micro TE cooler is to develop a reliable experimental method to determine contact resistivity on bulk TE materials. In the case of low resistance ohmic contacts, the fraction of the voltage drop across the interface is very small compared to the voltage drop across the bulk TE material. In addition, the distribution of the current density across the contact area and into the semiconductor is non-uniform. Correct extraction of contact resistivity is difficult, especially below $10^{-6}$ $\Omega$-cm$^2$. There has been a considerable amount of work done on the characterization of ohmic contacts to Si and GaAs [3-10], but not much work has been reported on the measurement of metal contacts to bulk $Bi_2Te_3$-based TE materials.

To quantify the electrical contact resistance on bulk $Bi_2Te_3$ TE material, the authors have devised a device-like structure (DLS) utilizing numerous layers of TE material soldered together in a single reflow process and subsequently diced as shown in Figure 1. The resulting test device is measured for total electrical resistance. After precisely measuring the dimensions of the TE layers, solder layers and copper end-caps, the electrical contact resistance can be calculated. The advantages to this method are plentiful. Firstly, this testing technique allows measurements on bulk TE materials. Secondly, since the measurements are made on a DLS and include similar fabrication steps as a TE cooler, the resulting contact resistance should be representative of an actual TE cooler. And finally, the DLS can be cross-sectioned and inspected thoroughly to determine the actual thickness of each layer of the structure, thus minimizing errors due to the uncertainties in the structure geometries.

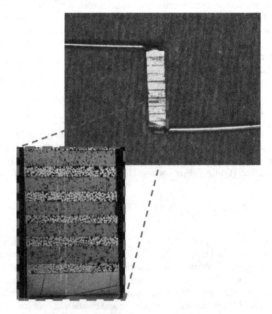

Figure 1: Device like structure (DLS) ready for testing

## Contact Resistance Model

The total electrical resistance of the DLS, $R_{total}$, is a function of the number of TE layers, $N$, and the thicknesses and electrical resistivities of the TE, solder and copper layers as shown in Equation 2.

$$R_{total} = \frac{\rho_{Cu}}{W_1 W_2} \sum_{i=1}^{2} t_{Cu,i} + \frac{\rho_{solder}}{W_1 W_2} \sum_{i=1}^{N+1} t_{solder,i} + \frac{\rho_{TE}}{W_1 W_2} \sum_{i=1}^{N} t_{TE,i} + \frac{2N\rho_{cont}}{W_1 W_2} \tag{2}$$

where $t$ is the thickness of the respective material, $\rho_{TE}$ is the bulk electrical resistivity of the respective material, $W_1$ and $W_2$ are the lateral dimensions of the DLS where $W_1 * W_2$ represents the cross-sectional area.

Equation 2 was solved for contact resistance and shown in Equation 3

$$\rho_{cont} = \frac{1}{2N} \left[ W_1 W_2 R_{total} - \rho_{Cu} \sum_{i=1}^{N} t_{Cu,i} - \rho_{solder} \sum_{i=1}^{N} t_{solder,i} - \rho_{TE} \sum_{i=1}^{N} t_{TE,i} \right] \tag{3}$$

## Uncertainty Model

For the measured value of contact resistance to have any significance, the uncertainty associated with this measurement needs to be quantified. Since the thicknesses of the TE, solder, and copper layers can be easily measured, the largest source of measurement error is associated with the bulk resistivity of the TE material. Accounting only for the error in the electrical resistivity of the TE material, the uncertainty in the electrical contact resistance can be described by Equation 4.

$$u_{\rho_{cont}} = \sqrt{C_{\rho_{TE}}^2 \varepsilon_{\rho_{TE}}^2} \tag{4}$$

where $C_{\rho_{TE}}$ is the error coefficient for the bulk resistivity and $\varepsilon_{\rho_{TE}}$ is the error of the bulk resistivity of the TE material.

To estimate the error, $\varepsilon$ in the resistivity of the TE material several samples were cut from a single wafer of material, and the standard deviation of the sample population was calculated. If the error in the resistivity of the TE material is estimated to be equal to this standard deviation, $\sigma$ then Equation 4 becomes 5.

$$u_{\rho_{cont}} = \left| C_{\rho_{TE}} \sigma_{\rho_{TE}} \right| \tag{5}$$

To calculate the coefficient, C in this equation, the partial derivative of Equation 3 with respect to TE resistivity is used as seen in Equation 6.

$$C_{\rho_{TE}} = \frac{\partial \rho_{cont}}{\partial \rho_{TE}} = -\frac{1}{2N} \sum_{i=1}^{N} t_{TE,i} \tag{6}$$

Therefore the uncertainty of the measured contact resistance due solely to the variation in TE bulk resistivity is described by Equation 7.

$$u_{\rho_{cont}} = \frac{\sigma_{\rho_{TE}}}{2N} \sum_{i=1}^{N} t_{TE,i} \tag{7}$$

## EXPERIMENTAL DETAILS

As shown in Figure 2(a and b), n- and p-type $Bi_2Te_3$ wafers, 0.008" ± 0.0005" in thickness were segmented into strips. The first strip of $Bi_2Te_3$ alloy wafer was placed on a specially designed fixture. Solder was applied on it manually with a ceramic strip ensuring flatness and smoothness. Then another TE strip was placed on it and pressed down gently to force excess solder out. The similar process continued until all strips were stacked on top of each other as shown in Figure 2(c). 10 strips were stacked together and joined with tin and antimony based solder. Using a BTU Paragon 70 reflow furnace, the clamped stack was reflowed in a BTU Paragon 70 reflow furnace. Several test runs were performed to ensure uniform solder reflow and repeatability. After the reflow process, the stacked-up device was diced into 0.050" x 0.050" squares. As shown in Figure 1, a 0.010" thick, 0.050"wide Cu strips were cut into 1" long sections and attached to the ends of the 0.050"x 0.050" diced devices as lead wires for testing. Lead wired devices were used for total resistance measurement and testing. This method is based on measurement of total resistance across a stacked device, which accounts for resistance from TE material, contact resistance from interfaces, and resistance from solder and Cu tabs. A Keithley milliohm-meter 503 was employed for precise measurement of total resistance. The 503 Keithley milliohm meter, an AC device, has been calibrated to measure resistance as low as 1 $m\Omega$ accurately.

Figure 2: Process steps for to prepare device for contact resistance measurement a) single wafer b) diced wafer ready for stack-up and c) stack of 10 partial wafers after soldering and re-flowing

## DISCUSSION

The contact resistance measurements for metal contacts to n- and p-type Bi2Te3 are shown in Tables 1 and 2 respectively. The metals contacts are made using Marlow's proprietary process of etching and plating. Specific contact resistivity is extracted using Equations 2 and 3 after total resistance measurement of a DLS. Bulk resistivity values of 1.01 and 0.95 $m\Omega$-cm were used for calculations for n- and p-type, respectively.

Table 1: N-type standard process

| | 1 | 2 | 3 | 4 | 5 | 6 |
|---|---|---|---|---|---|---|
| Area ($\mu m^2$) | 1.68 | 1.69 | 1.73 | 1.63 | 1.66 | 1.65 |
| $R_{Cu}$ (m$\Omega$) | 0.01 | 0.01 | 0.01 | 0.01 | 0.01 | 0.01 |
| $R_{solder}$ (m$\Omega$) | 0.09 | 0.10 | 0.09 | 0.10 | 0.09 | 0.09 |
| $R_{TE}$ (m$\Omega$) | 10.95 | 10.53 | 10.64 | 11.25 | 10.73 | 10.92 |
| $R_{total}$ (m$\Omega$) | 12.50 | 12.00 | 12.50 | 12.75 | 12.25 | 12.50 |
| $R_{cont}$ (m$\Omega$) | 1.45 | 1.36 | 1.77 | 1.40 | 1.42 | 1.48 |
| $R_{cont}$ per interface ($\mu\Omega$) | 72.5 | 68.2 | 88.4 | 69.9 | 70.8 | 74.1 |
| $\rho_{cont}$ ($10^{-6} * \Omega$-cm$^2$) | 1.22 | 1.15 | 1.53 | 1.14 | 1.18 | 1.22 |

Table 2: P-type standard process

| | 1 | 2 | 3 | 4 | 5 | 6 | 7 | 8 |
|---|---|---|---|---|---|---|---|---|
| Area ($\mu m^2$) | 1.87 | 1.73 | 1.69 | 1.55 | 1.47 | 1.6 | 1.55 | 1.61 |
| $R_{Cu}$ (m$\Omega$) | 0.01 | 0.01 | 0.01 | 0.01 | 0.01 | 0.01 | 0.01 | 0.01 |
| $R_{solder}$ (m$\Omega$) | 0.08 | 0.08 | 0.08 | 0.1 | 0.1 | 0.09 | 0.09 | 0.09 |
| $R_{TE}$ (m$\Omega$) | 9.84 | 10.64 | 10.86 | 11.8 | 12.37 | 11.54 | 12.44 | 10.77 |
| $R_{total}$ (m$\Omega$) | 10.50 | 12.00 | 12.00 | 13.50 | 13.25 | 13.00 | 14.50 | 11.25 |
| $R_{cont}$ (m$\Omega$) | 0.58 | 1.27 | 1.05 | 1.60 | 0.77 | 1.36 | 1.96 | 0.39 |
| $R_{cont}$ per interface ($\mu\Omega$) | 29.1 | 63.4 | 52.5 | 79.8 | 38.5 | 68.1 | 98.1 | 19.3 |
| $\rho_{cont}$ ($10^{-6} * \Omega$-cm$^2$) | 0.54 | 1.09 | 0.89 | 1.24 | 0.56 | 1.09 | 1.52 | 0.31 |

Table 3 summarizes the average of 6 devices for n-type and an average of 8 devices for p-type Bi$_2$Te$_3$. Specific contact resistivity for n-type was measured to be 1.2 ± 0.14 $\mu\Omega$-cm$^2$. Specific contact resistivity for p-type was measured to be 0.9 ± 0.4 $\mu\Omega$-cm$^2$.

Table 3: Summary of contact resistance measurement on n- and p-type

| | n-type | p-type |
|---|---|---|
| Contact Resistance Average ($10^{-6} * \Omega$-cm$^2$) | 1.24 | 0.91 |
| Standard deviation ($10^{-6} * \Omega$-cm$^2$) | 0.14 | 0.41 |

The results from Tables 1 and 2 assume that the electrical resistivities of the bulk TE materials are well known. But for production Bi$_2$Te$_3$ materials, there are established variations between ingots, along the same ingot, and even within a single slice of an ingot, resulting in uncertainty of the bulk resistivity for these test measurements. Using Equation 7 and the measured average TE material thickness of 0.008 inches, the uncertainties associated with the measured contact resistances are calculated for a range of TE material resistivity standard deviations and given in Table 4.

Table 4: Uncertainty in contact resistance due to electrical resistivity of bulk TE material

| $\sigma_{pTE}$ ($10^{-3} * \Omega$-cm) | $u_{pcont}$ ($10^{-6} * \Omega$-cm$^2$) |
|---|---|
| 1.00E-04 | 0.000635 |
| 1.00E-03 | 0.00635 |
| 1.00E-02 | 0.0635 |
| 1.00E-01 | 0.635 |

It is clear that as the standard deviation of the measured bulk TE material resistivity increases, the resulting uncertainty in the measured contact resistance also increases. For these experiments, the standard deviation of the bulk TE material resistivity for n-type was measured to be $1 \times 10^{-02}$ m$\Omega$-cm, contributing to the uncertainty of the contact resistance by $6 \times 10^{-02}$ $\mu\Omega$-cm$^2$. This suggests that the bulk resistivity of the TE material had relatively little contribution to the overall variation in results for n-type. But for p-type, the standard deviation of bulk TE material resistivity was measured to be $6 \times 10^{-02}$ m$\Omega$-cm contributing an uncertainty of $3.6 \times 10^{-1}$ $\mu\Omega$-cm$^2$, which explains the higher standard deviation of $4 \times 10^{-1}$ $\mu\Omega$-cm$^2$ for specific contact resistivity (see Table 3).

## CONCLUSIONS

A new approach to measurement of contact resistance is developed that can precisely measure contact resistance (on the order of $5 \times 10^{-01}$ $\mu\Omega$-cm$^2$) on bulk thermoelectric materials by processing stacks of bulk, metal-coated TE wafers using TE industry standard processes. This technique is shown to exploit realistic TE device manufacturing techniques and results in an almost device-like structure, therefore representing a realistic value for electrical contact resistance in a bulk TE device. The specific contact resistivity of metal contacts to bulk TE material was measured to be $1.2 \pm 0.14$ $\mu\Omega$-cm$^2$ for n-type and $0.9 \pm 0.4$ $\mu\Omega$-cm$^2$ for p-type. The standard deviations and therefore the uncertainties in the measured contact resistivities are likely strongly influenced by the variation in bulk TE material resistivity in the case of the p-type alloy, but not in the case of the n-type alloy.

## ACKNOWLEDGMENTS

The authors would like to thank Sal Rumy for working on optimization of DLS fabrication process and John White for useful discussions.

## REFERENCES

[1]    H.J. Goldsmid, Electronic Refrigeration, pp. 9, Pion Limited, London (1986).
[2]    A.S. Arnold, J.S. Wilson, and M.G. Boshier, Review of Scientific Instruments, **69**, 1236 (1998).
[3]    S.L. Zhang, Microelectronic Engineering, **70**, 174 (2003).

[4]     V. Narayanan, Z. Liu, Y.M.N. Shen, M. Kim, and E.C. Kan, IEDM Tech. Digest, **87** (2000).

[5]     S.P. Zimin, V.S. Kuznetsov, and A.V. Prokaznikov, Applied Surface Science, **91**, 355 (1995).

[6]     K. Yamada, K. Tomita, and T. Ohmi, Applied Physics Letters, **64**, 3449 (1994).

[7]     Z. Z. Chen, Z.X. Qin, Y.Z. Tong, X.D. Hu, T.J. Yu, Z.J. Yang, X.M. Ding, Z.H. Li, and G.Y. Zhang, Material Science and Engineering B, **100**, 199 (2003).

[8]     A. Katz, S. Nakahara, W. Savin, and B. E. Weir, Journal of Applied Physics, **68**, 4133 (1990).

[9]     M. O. Aboelfotoh, C. L. Lin, and J. M. Woodall, Applied Physics Letters, **65**, 3245 (1994).

[10]    R.P. Gupta, K. Xiong, J.B. White, Kyeongjae Cho, H.N. Alshareef, and B.E. Gnade, Journal of The Electrochemical Society, **157**(6), H666 (2010).

Mater. Res. Soc. Symp. Proc. Vol. 1490 © 2013 Materials Research Society
DOI: 10.1557/opl.2013.26

# Investigation of Wide Bandgap Semiconductors for Thermoelectric Applications

*B. Kucukgok[1], Q. He[1]; A. Carlson[3], A. G. Melton[1], I. T. Ferguson[1] and N. Lu[1, 2,a]*

[1]Department of Electrical and Computer Engineering, University of North Carolina at Charlotte, 9201 University City Blvd, Charlotte, NC 28223, U.S.A.

[2]Department of Engineering Technology, University of North Carolina at Charlotte, 9201 University City Blvd, Charlotte, NC 28223, U.S.A.

[3]Department of Mechanical Engineering and Engineering Science, University of North Carolina at Charlotte, 9201 University City Blvd, Charlotte, NC 28223, U.S.A.

## ABSTRACT

Thermoelectric materials with stable mechanical and chemical properties at high temperature are required for power generation applications. For example, gas temperatures up to $1000^\circ C$ are normally present in the waste stream of industrial processes and this can be used for electricity generation. There are few semiconductor materials that can operate effectively at these high temperatures. One solution may be the use of wide bandgap materials, and in particular GaN-based materials, which may offer a traditional semiconductor solution for high temperatures thermoelectric power generation. In particular, the ability to both grow GaN-based materials and fabricate them into devices is well understood if their thermoelectric properties are favorable. To investigate the possibility of using III-Nitride and its alloys for thermoelectric applications, we synthesized and characterized room temperature thermoelectric properties of metal organic chemical vapor deposition grown GaN and InGaN with different carrier concentrations and indium compositions. The promising value of Seebeck coefficients and power factors of Si-doped GaN and InGaN indicated that these materials are suitable for thermoelectric applications.

## INTRODUCTION

The concept of obtaining electrical energy from waste heat energy has received great attention by many researchers in recent years. This is because thermoelectric (TE) energy conversion is efficient and adaptive to a variety of applications, such as automobiles' engine coolant or exhaust gas, and diesel power plants [1]. In addition, TE generators are quite suitable for power generation and energy harvesting applications [2].

The performance of a thermoelectric material is determined by its dimensionless figure-of-merit (ZT) and expressed as:

$$ZT= S^2\sigma T/\kappa_e + \kappa_L \qquad (1)$$

where S, $\sigma$, $\kappa_e$, $\kappa_L$, and T are the Seebeck coefficient, electrical conductivity, electrical and lattice thermal conductivities, and absolute temperature respectively. The power factor of a thermoelectric material is given by $P= S^2\sigma$. To achieve high efficiency thermoelectric conversion a high ZT value needs to be obtained. Two approaches can be taken to achieve this end. The first approach is to pursue a high S value, which requires wide-bandgap energy, high charge carrier effective masses [3], and high electrical conductivity. The second approach is to reduce the thermal conductivity of the material [4]. Increasing phonon scattering is especially important to achieving low thermal conductivity, since phonon transport is the largest component of total thermal conductivity in semiconductors [4,5]. Non-traditional crystal structures can enable the control of both phonon scattering and electrical conductivity [6,7]. Such materials are designed with the 'phonon-glass electron crystal' principle, which seeks to minimize electron scattering (as in a crystal) and maximize phonon scattering (as in a glass) in order to achieve high ZT [4,6,7].

Good thermoelectric materials are often doped semiconductors, since insulators have low electrical conductivity, and metals have low Seebeck coefficient and high thermal conductivity [8,9]. Recent studies have shown that wide-bandgap semiconductor, such as III-Nitride alloys, are promising materials for next-generation high efficiency TE applications. Several groups have studied TE effects in Nitride-based semiconductors, both theoretically and experimentally [10-16]. III-Nitrides have several advantageous properties for TE applications, such as high temperature stability, high mechanical strength, and radiation hardness [12]. In addition, alloying and nanostructuring are easy to implement in nitride materials, which enable tuning the thermoelectric properties [12,13].

In this study, we investigated the thermoelectric properties of Si-doped GaN and InGaN alloys with various doping concentration and alloy content. The n-type GaN and $In_xGa_{1-x}N$ thin films were grown by metal organic chemical vapor deposition (MOCVD).

**EXPERIMENT**

All the materials were grown by MOCVD on a c-plane (0001) sapphire substrate. N-type (silicon-doped) GaN films were grown using trimethylgallium [TMGa, $Ga(CH_3)_3$], ammonia ($NH_3$), and silane ($SiH_4$) as precursors. InGaN films were grown using trimethylindium [TMIn, In $(CH_3)_3$], triethylgallium [TEGa, $Ga(C_2H_5)_3$], and $NH_3$ as the gas-phase precursors. P-type (magnesium-doped) InGaN films were also grown using a $Cp_2Mg$ precursor. The growth procedure and epitaxial layer thickness were monitored in-situ, using an optical reflectometer. InGaN layers of various compositions (11% to 16%) and thicknesses (80 nm to 250 nm) were grown. A list of samples and their electrical properties and Seebeck coefficients were summarized in Table.1.

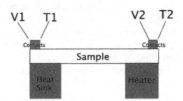

**Figure 1.** Schematic illustration of custom Seebeck measurement setup. V1-2 and T1-2 are voltage probes and thermocouples respectively.

Table.1 A list of samples and their electrical properties and Seebeck coefficients at room temperature, 300K. All the samples' dimensions used in this study are ≈12mm length, and ≈4mm width.

| Sample | Seebeck Coefficient ($\mu$V/K) | Power Factor ($10^{-4}$W/mK$^2$) | Mobility (cm$^2$V$^{-1}$s$^{-1}$) | Conductivity ($\Omega^{-1}$cm$^{-1}$) | Carrier Density (cm$^{-3}$) |
|---|---|---|---|---|---|
| In$_{0.11}$Ga$_{0.89}$N | -525 | 6.36 | 268 | 1.37 | 3×16 |
| In$_{0.12}$Ga$_{0.88}$N | -255 | 1.26 | 10 | 5.15 | 3×17 |
| In$_{0.14}$Ga$_{0.86}$N | -215 | 5.73 | 10 | 8.06 | 5×18 |
| In$_{0.16}$Ga$_{0.84}$N | -335 | 1.62 | 750 | 145.11 | 1×18 |
| GaN:Si | -224 | 4.51 | 480 | 90.04 | 1×18 |
| GaN:Si | -90 | 9.67 | 254 | 119.51 | 2×18 |
| GaN:Si | -115 | 2.72 | 245 | 205.91 | 5×18 |
| GaN:Si | -99 | 1.55 | 173 | 158.71 | 6×18 |
| GaN:Si | -105 | 1.98 | 190 | 180.01 | 6×18 |

The thermal gradient method was used to determine Seebeck coefficients and a custom instrument was used for Seebeck measurements as depicted in Figure 1 [10,12,13,15,16]. Room temperature in-plane Seebeck measurements were performed by placing the sample across two commercial thermoelectric modules which one module was heating one end of the sample while other one, on the other end of the sample, was held at constant temperature during measurements. As thermal gradient was generated in the sample, Seebeck voltage was measured across the sample using voltage probe system via indium (In) contacts, which was placed to the top surface of sample and temperature on each side of the sample was measured using type K thermocouples. The Seebeck coefficients were determined by the slope of a graph, Seebeck voltage gradient as a function of temperature gradient across the sample, and Seebeck voltages for all samples showed perfect linearity ($R^2 \approx 0.999$) with the measured temperature gradient. Electrical conductivity measurements were performed using standard Van der Pauw Hall-effect measurement.

## DISCUSSION

Figure 2 (a) and (b) exhibit Seebeck coefficient and power factor as a function of carrier density of GaN:Si materials. Figure 2 (a) shows that seebeck coefficient has decreased with the increase of carrier concentration, although the carrier concentrations for Si doped samples are within the same order of magnitude. The interrelationship between carrier density and the Seebeck coefficient is expressed as:

$$\alpha = \frac{8\pi^2 k_b^2}{3eh^2} m*T\left(\frac{\pi}{3n}\right)^{2/3}$$

(2)

where $k_B$, h, m$^*$, e, T, and n are Boltzmann constant, plank's constant effective mass, electric charge, absolute temperature, and carrier density, respectively. Reward dopant concentration is inversely correlated to the Seebeck coefficient. A negative Seebeck coefficient indicates that all the samples have n type conductivity. Figure 2(b) depicts the power factor of measured GaN:Si as a function of carrier density. The result indicates that the optimized carrier concentration is $2\times10^{18}$ cm$^{-3}$ to achieve maximum power factor, due to an inverse relationship between electrical conductivity and the Seebeck coefficient. In addition, as equation (2) indicates that effective mass of the charge carriers also have an impact on Seebeck coefficient.

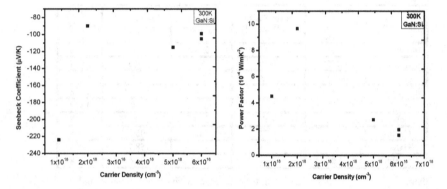

**Figure 2.** (a) Seebeck coefficient vs. carrier density of n-type GaN (b) Power factor vs. carrier density of n-type GaN

Figure 3 (a) shows measured seebeck coefficient of $In_xGa_{1-x}N$ alloys as functions of In content ($0.11 \leq x \leq 0.16$). Adding In into GaN reveals simultaneous drop in the bandgap of the material, thus reduces the Seebeck coefficient whereas lowering In composition results in high Seebeck values. In addition, increased background electron concentration with introducing more In alloys induces tradeoff between Seebeck coefficient and electrical conductivity [12]. One reason could be electrical conductivity is increasing with adding more indium content due to an increase in electrically active crystal defects, usually acting as donors [15]. In Figure 3 (b) we plot the measured power factor values for $In_xGa_{1-x}N$ alloys with various In compositions at 300 K and except $In_{0.12}Ga_{0.88}N$ one can suggest that power factor is decreasing while x is increasing. One reason could be electrical conductivity of InGaN is reducing with mobility as shown in Figure 4, since introducing In alloys into InGaN increase alloy scattering effects [13].

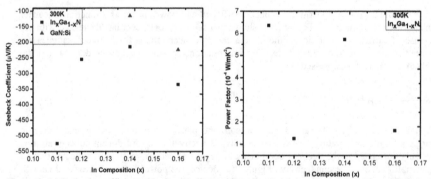

**Figure 3.** (a) Seebeck coefficient of $In_xGa_{1-x}N$ alloys with various In compositions (x) (b) The measured power factors for $In_xGa_{1-x}N$ alloys with various In compositions (x) at 300 K

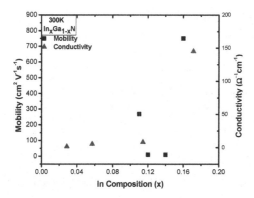

**Figure 4.** The measured electron mobility and conductivity for $In_xGa_{1-x}N$ alloys with various In compositions (x) at 300 K

## CONCLUSIONS

In conclusion, the thermoelectric properties of Si-doped GaN and InGaN alloys were investigated. Although Si doped GaN materials have doping concentration of same order of magnitude $\sim10^{18}$ $cm^{-3}$ trends of high Seebeck coefficients with the decrease of carrier concentration were observed, whereas a reduced power factor due to an inverse relationship of electrical conductivity of Seebeck coefficient.

The InGaN alloy with lower Indium content (x≈0.11) has shown higher Seebeck coefficient compared to higher Indium contents (x≈ 0.12, 0.14, and 0.16), since adding more indium into GaN decreases the bandgap while increasing electrical conductivity. Based on the reported experimental results, the GaN:Si and $In_{0.11}Ga_{0.89}N$ yielded very promising results of Seebeck coefficient and power factor of -224μV/K, $9.67\times10^{-4}W/mK^2$ and -525μV/K, 6.36 $W/mK^2$ respectively. Uncertainty of Seebeck measurements was estimated at ± 5% due to vibration and pressure on the contacts caused by voltage probes. III-Nitride materials play an important role for thermoelectric application with the further improvement of thermoelectric properties through bandgap engineering and nanostructuring.

## REFERENCES

1. J.W. Fergus, Journal of the European Ceramic Society **32**, 525-540 (2012).
2. C. Hadjistassou, E. Kyriakides, J. Georgiou, Energy Conversion and Management **66**, 165-172 (2013).
3. A. Kosuga, T. Plirdpring, R. Higashine, M. Matsuzawa, K. Kurosaki, and S. Yamanaka, Applied Physics Letters **100**, 042108 (2012).
4. S.H. Lo, J. He, K. Biswas, M.G. Kanatzidis, and V. P. Dravid, Advanced Functional Materials, (2012).
5. O. Delaire1, J. Ma, K. Marty, A. F. May, M. A. McGuire, M-H. Du, D. J. Singh, A. Podlesnyak, G. Ehlers, M. D. Lumsden and B. C. Sales, Nature **10**, (2011).

6. H. Liu, X. Shi, F. Xu, L. Zhang, W. Zhang, L. Chen, Q. Li, C. Uher, T. Day and G. J. Snyder, Nature 11, (2012).

7. G.J. Snyder, E.S. Toberer, Nature 7, (2008).
8. A. Shakouri, Annual Review Materials Research 41, 399-431 (2011).

9. T.M Tritt, Annual Review Material Research, 41, 433-448 (2011).
10. W. Liu and A.A. Balandin, Journal of Applied Physics 97, 073710 (2005).
11. S. Yamaguchi, R.Izaki, N.Kaiwa, A.Yamamoto, Applied Physics Letters 86, 252102, (2005).
12. B.N. Pantha, R. Dahal, J.Li, J.Y. Lin, H.X. Jiang, Applied Physics Letters 92, 042112 (2008).
13. A. Sztein, H. Ohta, J. Sonoda, A. Ramu, J. E. Bowers, S. P. DenBaars, and S. Nakamura, Applied Physics Express 2, 111003 (2009).

14. H. Tong, H. Zhao, V.A. Handara, J.A. Herbsommer and N. Tansu, SPIE 7211, 721103 (2009).

15. E. N. Hurwitz, B. Kucukgok, A. G. Melton, Z. Liu, N. Lu, I. T. Ferguson, Material Research Society Symposium Proceeding, 1396, (2012).
16. A. Sztein, H. Ohta, J. E. Bowers, S. P. DenBaars, and S. Nakamura, Journal of Applied Physics 110, 123709 (2011).

Mater. Res. Soc. Symp. Proc. Vol. 1490 © 2012 Materials Research Society
DOI: 10.1557/opl.2012.1572

# Nano-scale vacuum spaced thermo-tunnel devices for energy harvesting applications

Amit K. Tiwari, Jonathan P. Goss, Nick G. Wright, and Alton B. Horsfall

Electrical and Electronic Engineering, Newcastle University, NE1 7RU, UK

## ABSTRACT

The output power-density and the efficiency of thermo-tunnel devices are examined as a function of inter-electrode separation, electrode work-function, and temperature. We find that these physical parameters dramatically influence the device characteristics, and under optimal conditions a thermo-tunnel device is capable of delivering a very high output power-density of $\sim 10^3\,\mathrm{Wcm^{-2}}$. In addition, at higher temperatures, the heat-conversion efficiency of the thermo-tunnel device approaches $\sim 10\%$, comparable to that of a thermoelectric generator. We therefore propose that thermo-tunnel devices are promising for solid-state thermal energy conversion.

## INTRODUCTION

Solid-state thermal energy conversion technologies are becoming increasingly important for both terrestrial and extra-terrestrial applications. Thermionic energy-conversion is of particular interest because of the potential of providing relatively efficient thermal-energy conversion in comparison to commonly used techniques, such as thermoelectric and thermophotovoltaic conversion [1]. Recently, combined thermionic emission and tunneling of hot electrons (thermo-tunneling) has come to the fore, capable of providing very high output power-densities via a nano-scale, vacuum-spaced thermo-tunnel device [2–10]. The height and the width of a potential barrier between the electrodes which an electron must overcome are controlled by the physical parameters, particularly interelectrode separation ($d$) and the work function of the electrodes ($\phi_e$ and $\phi_c$ for the emitting and collecting electrodes) [2–6], with both the engineering of low work-functions and uniform nano-meter separations over relatively large areas being simultaneous challenges to overcome. Recent advances in work function reduction of semiconductor surfaces, particularly of diamond [11, 12], and in parallel, the fabrication of nano-scale vacuum gap [8–10, 13] show that fabrication of such free-standing structures will be possible in near future. By applying a thermal gradient across the ultra-thin (a few nm) vacuum gap of a tunnel device, a large unidirectional flux of thermionic and thermo-tunnel currents between an emitter and collector, and a thermal potential difference can be achieved.

Thermo-tunnel devices offer several advantages over conventional thermionic and thermoelectric devices. Since $d$ in a thermo-tunnel device is extremely small, the negative-space-charge effects [1], which impede the flow of electrons from emitter to collector in a conventional thermionic converter, are negligible. For thermoelectric devices, in addition to a high Seebeck coefficient, high electrical but low thermal conductivity are necessary to achieve a high figure of merit [1]. In contrast, thermo-tunnel devices have no such requirement, although in practice the encapsulating material must have a low thermal conductivity to prevent a thermal short between the electrodes.

For an optimal $d$ and temperature gradient, performance of a thermo-tunnel device principally depends upon the electrode workfunctions, which can be modified in case of negative electron affinity materials, which is the case for some specific diamond terminations [6, 11, 12].

A number of models, mostly for the cooling applications of the thermo-tunnel effect, have been presented in the past, which demonstrate that at a small $d$, a significant output power-density in the range of 10–1000 Wcm$^{-2}$ is achievable[4, 6, 9]. However, for energy scavenging applications, the thermo-tunnel effect has received relatively little attention to-date[5, 13]. For the fabrication of an efficient power-generating thermo-tunnel device, optimization of geometrical and physical parameters, such as temperatures, work functions, and $d$, is understood to be a prerequisite. We have

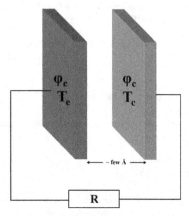

Figure 1: Schematic of a thermo-tunnel device. The left electrode (red) is the emitter, and the right (blue) the collector.

therefore investigated the functioning and optimized physical parameters of a thermo-tunnel device for efficient thermal energy conversion.

## THEORY

To calculate the net thermo-tunnel (combined thermionic and tunneling) current between the emitter and the collector of a thermo-tunnel device (Fig. 1), it is necessary to obtain the interelectrode electric potential profile. For an electron moving from the emitter to the collector, the potential energy equation can be written as

$$V_e(x_e) = \phi_e - \frac{x_e}{d}\left(eV_{\text{bias}} + \phi_e - \phi_c\right) - \frac{e^2}{4\pi\varepsilon_0}\left[\frac{1}{4x_e} + \frac{1}{2}\sum_{n=1}^{\infty}\left(\frac{nd}{n^2d^2 - x_e^2} - \frac{1}{nd}\right)\right], \quad (1)$$

where $x_e$ is the distance from the emitter into the vacuum. The first two terms taken into account the effect of the bias ($V_{\text{bias}}$) and the work functions of the emitter ($\phi_e$) and the collector ($\phi_c$), while the third term corresponds to the image charge effect. The $V_{\text{bias}}$ is the difference between the electron chemical potentials of the emitter ($E_{fe}$) and the collector ($E_{fc}$), which strongly depends upon the temperature.

It is important to note that $d$ has a strong impact upon both the magnitude and the shape of the interelectrode potential barrier. Fig. 2(a) shows that the barrier height decreases rapidly with decreasing $d$ and the potential barrier becomes more strongly peaked. This indicates that for a small value of $d$, even a low energy electron can tunnel through the barrier. On increasing the work function of the emitter, the potential barrier, Fig. 2(b), becomes asymmetric and the barrier height increases continuously.

Reduction in the width and the height of a potential barrier decreases the minimum kinetic energy (KE) required for the electrons to be emitted into the vacuum. It is evident from Fig. 3 that for an easily overcome potential barrier, even low KE electrons have a relatively high probability of tunneling through the barrier. Using the WKB

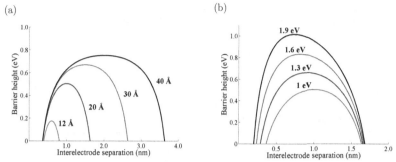

(a)                                    (b)

Figure 2: Potential barriers for an electron moving from the emitter to collector for (a) different values of $d$ with $\phi_e = \phi_c = 1\,\text{eV}$, and (b) different values of $\phi_e$ with $d = 20\text{Å}$ and $\phi_c = 1\,\text{eV}$.

approximation, the tunneling probability for the emitter electrons can be expressed as

$$D_e(E_x) = \exp\left[-\frac{2}{\hbar}\int_{x_{e1}}^{x_{e2}}\sqrt{2m\left[V_e(x_e) - E_x\right]dx_e}\right] \quad \text{if} \quad E_x < V_{e,\text{max}}$$

$$= 1 \quad \text{otherwise} \tag{2}$$

where $m$ is the mass of an electron. In eq. 2, $x_{e1}$ and $x_{e2}$ are the roots of the equation

$$V_{e,\text{max}} - E_x = 0. \tag{3}$$

The total current density, a combination of thermionic and tunneling current-densities, can be written analytically as

$$J_{e,\text{tot}} = e\int_{-\infty}^{\infty} N_e(E_x)D_e(E_x)dE_x \tag{4}$$

Hence, the tunneling and thermionic contributions, $J_{e,\text{tun}}$ and $J_{e,\text{ther}}$, can be separated into

$$J_{e,\text{tun}} = e\int_{-\infty}^{V_{e,\text{max}}} N_e(E_x)D_e(E_x)dE_x \quad \text{and} \quad J_{e,\text{ther}} = e\int_{V_{e,\text{max}}}^{\infty} N_e(E_x)D_e(E_x)dE_x. \tag{5}$$

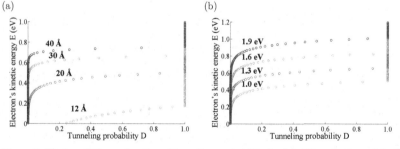

(a)                                    (b)

Figure 3: Electron KE as a function of tunneling probability for (a) different values of $d$ with $\phi_e = \phi_c = 1\,\text{eV}$, and (b) different values of $\phi_e$ with $d = 20\text{Å}$ and $\phi_c = 1\,\text{eV}$.

where $N_e(E_x)$ is the number of electrons per unit area in unit time that can escape from the (hot) emitter and reach the (cold) collector with their KE in $x$-direction in the range $E_x$, $E_x + dE_x$. $N_e(E_x)$ is calculated using

$$N_e(E_x) = \frac{4\pi m k_B T_e}{h^3} \ln\left[1 + \exp\left(\frac{-E_x}{k_B T_e}\right)\right], \tag{6}$$

where $k_B T_e$ is the average kinetic energy of electrons crossing the gap from the emitter to the collector.

Emitted electrons carry energy from the emitter to the collector, in a process known as emitter-electron cooling. The thermal power-density or emitter electron cooling can be calculated from

$$Q_e = \int_{-\infty}^{\infty} (k_B T_e + E_x) N_e(E_x) D_e(E_x) dE_x. \tag{7}$$

In a similar way, one can calculate the total thermo-tunnel current ($J_{c,tun}$ and $J_{c,ther}$) and thermal power density ($Q_c$) for the electrons emitted from the collector. The net thermo-tunnel current ($J_{tot}$) and thermal power density ($Q_{tot}$) can be expressed as

$$J_{tot} = J_{e,tot} - J_{c,tot} \quad \text{and} \quad Q_{tot} = Q_e - Q_c. \tag{8}$$

Using these relationships, it is possible to determine the operation of a thermo-tunnel device.

## RESULTS

Typical behavior of a thermo-tunnel device is shown in Fig. 4. Depending upon the bias, a thermo-tunnel device can be operated in three modes: heating, generating, and cooling. When the product of $J_{tot}$ and $V_{bias}$ is negative, this represent the generating behavior of the thermo-tunnel device. In this paper, only the generating behavior is presented. Similar to other solid-state energy-conversion devices, such as solar cells and thermo-electric generators, the maximum power point (MPP) has been calculated [Fig. 4(a)]. The maximum voltage ($V_{max}$) and the current density ($J_{tot,max}$) can then be obtained, corresponding to MPP. The maximum output power-density and efficiency, $\eta$, for a thermo-tunnel device can be expressed as

$$P_{max} = J_{max} \times V_{max} \quad \text{and} \quad \eta_{max} = \frac{P_{max}}{Q_{tot}}. \tag{9}$$

The generator behavior of a thermo-tunnel device has been investigated for different values of $d$, emitter temperature and $\phi_e$. The impact of $d$ upon the output power density is clearly evident in Fig. 4(b). Although the magnitude of output power-density of a thermo-tunnel device decreases rapidly with increasing $d$, the range of voltages within the generating regime has also increased. For example, typical voltage values corresponding to the generating region of 10Å lie in the range $-4.5$–$0$ eV, whereas for 15Å, these values vary in between $-7.0$ and $0$ eV. In the case of increasing $\phi_e$, we have observed similar behavior to that seen for increasing $d$. It is important to mention here that the generating region of a thermo-tunnel device is highly sensitive to the temperature of the electrodes. We have therefore investigated the effect of the emitter temperature on the maximum output power density as well as on the efficiency.

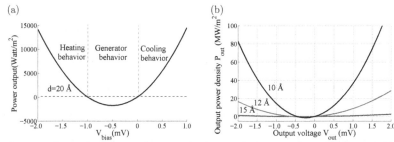

Figure 4: Power density as a function of voltage with $\phi_e = \phi_c = 1.0\,\text{eV}$, $T_e = 310\,\text{K}$ and $T_c = 300\,\text{K}$.

Efficiency and the output power density of a thermo-tunnel device as a function of emitter temperature are shown in Fig. 5(a). For a small values of $d$, such as 10 Å, the efficiency of a thermo-tunnel device has increased continuously from 0 to $\sim$1.3% in the temperature range from 301 to 700 K, which is equivalent to 57% of the Carnot efficiency. We note that the efficiency of a thermo-tunnel device increases very slowly with increasing $d$. However, on increasing the emitter temperature, a drastic change in the efficiency is observed. For example, the heat conversion efficiency corresponding to 23Å is calculated to be $\sim$11%.

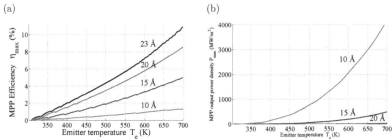

Figure 5: Dependence of (a) the efficiency and (b) the output power density on the emitter temperature when $\phi_e = \phi_c = 1\,\text{eV}$ and $T_c = 300\,\text{K}$.

The heat conversion efficiencies corresponding to small vacuum gaps and temperature gradients are found to be very small. We calculate that in contrast a previous study [5] the maximum efficiency of a thermo-tunnel device for a temperature interval of 10 K was 0.34% for $d = 25$ Å, which we have reproduced in our calculations. However, for the higher temperature difference 300 K, we obtain the much better value of around 8.5%, emphasising the importance of this operational parameter.

We further note that increases in $\phi_e$ have a relatively small impact upon the maximum voltage. However, due to the reduction in the output current-density, output power-density decreases rapidly. It has been suggested earlier [5] for efficient thermal energy conversion, the work functions of both the emitter and the collector should be minimized. This is indeed supported by our findings, which show that the maximum efficiency and power density of a thermo-tunnel device increases with decreasing work functions.

As shown in Fig. 5(b), due to very high thermo-tunnel current density at small $d$, the output power-density is predicted to be very large, which further increased with increasing emitter temperature. These results are consistent with Fig. 4, where the output power density is very small at large ineterelectrode separation.

## CONCLUSION

In conclusion, we have performed simulations to investigate the impact of physical parameters upon the generating behavior of a thermo-tunnel device. We find that upon increasing the inter-electrode separation and the emitter temperature, the efficiency of a thermo-tunnel device increases. On the contrary, output power-density rapidly increases with decreasing electrode separation. An increase in the work-function of the emitter has an adverse impact upon both the output power-density and efficiency. Results show that under optimized conditions, a large output power-density of around $10^3$ Wcm$^{-2}$, with efficiency in the range $\sim$ 8–12%, can be achieved via a thermo-tunnel device. It can therefore be concluded that for small values of electrode work-function and at moderately high emitter temperatures, nano-scale vacuum spaced thermo-tunnel devices are promising candidates for energy harvesting applications.

## ACKNOWLEDGMENT

This work is supported by BAE Systems and the Engineering and Physical Sciences Research Council (EPSRC) UK through the DHPA scheme.

## REFERENCES

[1] J. A. Angelo and D. Buden, *Space Nuclear Power* (Orbit Book Company, Malabar, Florida, 1985).

[2] Y. Hishinuma, T. H. Geballe, B. Y. Moyzhes, and T. W. Kenny, Appl. Phys. Lett. **78**, 2572 (2001).

[3] G. D. Mahan, J. Appl. Phys. **76**, 4362 (1994).

[4] T. L. Westover and T. S. Fisher, Phys. Rev. B **77**, 115426 (2008).

[5] G. Despesse and T. Jager, J. Appl. Phys. **96**, 5026 (2004).

[6] G. D. Mahan and L. M. Woods, Phys. Rev. Lett. **80**, 4016 (1998).

[7] A. N. Korotkov and K. K. Likharev, Appl. Phys. Lett. **75**, 2491 (1999).

[8] E. T. Enikov and T. Makansi, Nanotechnology **19**, 075703 (2008).

[9] Y. Hishinuma, T. H. Geballe, B. Y. Moyzhes, and T. W. Kenny, J. Appl. Phys. **94**, 4690 (2003).

[10] E.C. Teague, J. Res. Nat. Bur. Stand. **91**, 171 (1986).

[11] A. K. Tiwari, J. P. Goss, P. R. Briddon, N. G. Wright, A. B. Horsfall, and M. J. Rayson, Phys. Rev. B **86**, 155301 (2012).

[12] A. K. Tiwari, J. P. Goss, P. R. Briddon, N. G. Wright, A. B. Horsfall, R. Jones, and M. J. Rayson, unpublished, 2012.

[13] M. Arik, J. Bray, and S. Weaver, Nanosci. Nanotechnol. Lett. **2**, 189 (2002).

Mater. Res. Soc. Symp. Proc. Vol. 1490 © 2013 Materials Research Society
DOI: 10.1557/opl.2013.27

# Progress on Searching Optimal Thermal Spray Parameters for Magnesium Silicide

Gaosheng Fu[1], Lei Zuo*[1], Jon Longtin[1,] Yikai Chen[2] and Sanjay Sampath[2]
[1]Department of Mechanical Engineering, and [2]Department of Material Science and Engineering,
State University of New York at Stony Brook, Stony Brook, NY 11794, U.S.A.
(*lzuo@notes.cc.sunysb.edu, 631-632-9327)

ABSTRACT

The thermoelectric properties of $Mg_2Si$ coatings prepared by Atmospheric Plasma Spray (APS), and Vacuum Plasma Spray (VPS) are presented. Seebeck coefficient results of both APS and VPS have been reported. XRD and SEM analysis of the samples are also presented to understand how microstructure influences the coating thermoelectric properties. The results suggest significant improvements can be made on the reduction of impurity including oxidation and pure silicon by using proper spray method and parameters. Thermal spray has been demonstrated before to be effective way to reduce thermal conductivity which may due to the coating microstructure. VPS result shows higher Seebeck coefficient than APS which may due to lower level of oxidization.

Key words: thermoelectric material, $Mg_2Si$, thermal spray, APS, VPS

## 1. INTRODUCTION

Magnesium silicide ($Mg_2Si$) is particularly interesting at a temperature range from 400 to 800K as a thermoelectric solid-state power generator material [1, 2]. $Mg_2Si$ is abundant on earth, thus making it much less expensive than traditional Bismuth Telluride thermoelectric material. It is also non-toxic and friendly to the environment. The conversion efficiency of the thermoelectric material is determined by its figure of merit, $ZT=\sigma S^2 T/k$, where S is the Seebeck coefficient; $\sigma$ is the electrical conductivity; k is the thermal conductivity; and T is the absolute temperature.

Extensive research has been done to synthesize and solidify $Mg_2Si$ and its solid solutions, Zaitsev et al used direct melting $Mg_2Si$ and got a maximum ZT of 1.1[3]. Riffel and Schilz investigated mechanical alloying method followed by hot-pressing process [4]. Jung and Kim tried solid state reactions then hot press (HP) and obtained maximum ZT of 0.7 at 830K with Bi doped $Mg_2Si$ [5]. Kajikawa et al. [6, 7] and Tani[8] used spark plasma sintering and got maximum ZT of 0.86 at 862K with Bi doped $Mg_2Si$. The vertical Bridgman method has also been applied for the single crystal growth by Akasaka et al. [9] and Tamura et al. [10]. Melt-spinning followed by spark plasma synthesis (SPS) or hot pressing has been used to synthesis thermoelectric material by Q. Li and demonstrated thermoelectric properties could be enhanced by non-equilibrium technique [11].

As a flexible, industry-scalable and cost-effective manufacturing process, thermal spray has been traditionally used for protective barrier coating. In the past decade Sampath et al [12] extended it for material synthesis of functional electronics and sensors. Thermal spray has a very high quenching rate ($10^{6-7}$ K/sec) similar as the melt spinning [13], we would like to enable integrated manufacturing of thermoelectric devices on the exhaust system directly using thermal spray [14]. Thermal conductivity is reduced by thermal spray due to microstructure [15]. It should be noted that thermal spraying has already been used to prepare $FeSi_2$ as a thermoelectric material [16-19]. However, the maximum ZT of $FeSi_2$ is around 0.1 which is too low for practical energy recovery application. Thermal spray of $Mg_2Si$ is much more challenging and has not been reported by others, since Mg has much higher vapor pressure and higher reactivity with

O$_2$. In this study, thermal spraying technology, including Atmospheric Plasma Spray (APS) and Vacuum Plasma Spray (VPS) were applied to fabricate Mg$_2$Si coatings. Seebeck coefficient was characterized for thermal sprayed samples and compared with other methods including HP and SPS.

## 2. EXPERIMENT

Mg$_2$Si powder (purity > 98%) purchased from YHL New Energy CO.,Ltd (Zhejiang China) was used to deposit coatings on aluminum/titanium plates of size 40 x 25 x 2mm. The grain sizes of the powder range from 10 to 80 μm. The purer powders were not used because 1) starting from the low purity powder, proper thermal spray parameters can be found; 2) the purer Mg$_2$Si powder is very expensive and was not available with the required powder size and shape. For APS, Sulzer Metco F4-MB plasma torch and SG100 were used and for VPS the Plasma-Technik A-3000S VPS System is used. The parameters of the thermal sprays are shown in Table 1.

Table 1: Thermal spray parameters

| Sample | Temperature (°C) | Velocity (m/s) | Spray type | Other |
|--------|------------------|----------------|------------|-------|
| R1961 | 1781 | 155 | APS | 8mm nozzle |
| R1962 | 1834 | 183 | APS | 8mm nozzle |
| R2018 | 1850 | 155 | APS | 8mm nozzle, shroud |
| R2069 | 1820 | 232 | APS | 6mm nozzle |
| B006 | Low | High | APS | SG100 |
| VPS003 | N/A | N/A | VPS | 37kW power |
| VPS004 | N/A | N/A | VPS | 21kW power |
| VPS101 | N/A | N/A | VPS | 21kW, low substrate temp. |

The sample phase was analyzed by X-ray diffraction (XRD) (PAD-V, SCITAG Ing., CA, USA) utilizing Cu Kα radiation. The composition is calculated using MDI Jade software. Coating images were observed using a Scanning electric microscope (SEM) (Leo-1550-SFEG, Carl Zeiss SMT Ltd, Cambridge UK). Seebeck coefficient is measured using MMR's SB-100 Seebeck Measurement System and K20 temperature controller (MMR Technologies, CA). The substrate of the sprayed coating is polished and a 1x1x4mm long bar free standing sample is used for Seebeck measurement.

## 3. RESULTS AND DISCUSSION

### 3.1 Phase composition of thermal sprayed coatings

The XRD results of the APS are compared with the feed stock powder which is shown in in Figure 1(a). As can be seen, thermal spray results in oxidation, which is generally detrimental to thermoelectric properties. Samples R1961 and R1962 are both APS with 8mm nozzle, but R1962 is in higher temperature and velocity, which gave higher concentration of Si. The reaction between Mg$_2$Si and O$_2$ is around 450 °C, while the temperature during thermal spray is significantly higher than this. The reaction is:

$$450 \text{ °C}$$
$$Mg_2Si + O_2 \quad \text{-----}> \quad MgO + Si$$

Several techniques have been tried to reduce the oxidation of Mg$_2$Si during the spray process, including APS with a shroud (R2018), APS with a smaller nozzle (R2069) and APS with an SG100 gun (B006). From table 2, we see APS with shroud does help but the improvement is limited. APS with smaller nozzle (R2069) has better result but oxidation is still too high for thermoelectric applications. We tried another set of high velocity low temperature APS using SG100 gun with internal powder passage to avoid oxidation. The XRD result of B006 is very consistent with the feedstock powder, although very small MgO and Si peak are found. The reason B006 shows Mg peak is low temperature and high velocity condition which suppress the oxidation of Mg. Meanwhile, the internal passage of powder in SG100 gun can further eliminate the chance of reaction between powder and ambient air.

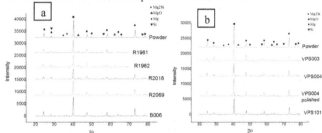

Figure 1: XRD of APS (a) and VPS (b) coating

In attempt to reduce the oxidation during the spray, VPS was used to prepare Mg₂Si samples. Figure 1(b) shows the XRD results for VPS. Improvement can be seen, that the lower O2 concentration helped reduce oxidation in APS. Normally, the VPS has higher temperature than APS because of lower heat convection in the vacuum chamber. Mg has a very high vapor pressure, so high temperate in VPS can evaporate Mg and leave Si behind. This is why VPS coatings have higher concentration of Si. It appears to be difficult to completely avoid oxidation and Si by VPS. VPS101 is the most recent coating fabricated using 21kW and substrate temperature control method.

Table 2: Concentrations of different composition in powder and coatings

|  | Mg₂Si | MgO | Si | Mg |
|---|---|---|---|---|
| Powder003 | 98.3% | 1.3% | - | 0.4% |
| R1961 | 77.5% | 17.0% | 5.5% | - |
| R1962 | 76.0% | 16.7% | 7.3% | - |
| R2018 | 78.6% | 14.9% | 6.5% | - |
| R2069 | 86.0% | 10.4% | 3.6% | - |
| B006 | 91.5% | 3.0% | - | 5.5% |
| VPS003 | 74.7% | 6.2% | 19.1% | - |
| VPS004 | 88.3% | 5.8% | 5.9% | - |
| VPS101 | 89.2% | 5.8% | 5.0% | - |

From figure 1 and Table 2, VPS did reduce some oxidation but the concentration is still far too high for practical applications. Part of the issue may be the high reactivity and vapor pressure of Mg. The vacuum produced during VPS is around $10^{-4}$ Torr hence a small quantity of oxygen will still be present that can react with Mg at high processing temperatures. More investigation of VPS for Mg₂Si is needed to decide whether it is a viable approach for thermal spraying of Mg₂Si materials for thermoelectric applications.

## 3.2 SEM result of thermal sprayed coatings

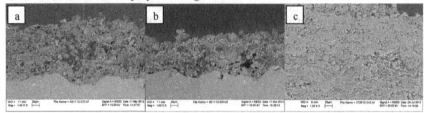

Figure 2: SEM of APS samples: a) R1961, b) R1962, c) B006

Figure 2 shows SEM images of several APS samples. The low temperature high speed APS (B006) gives a denser microstructure, while bonding in R1961 and R1962 is quite poor. B006 shows softening instead of melting; this condition just softens the powder.

Figure 3 shows the higher temperature results in a denser coating. VPS003 and VPS004 were sprayed at 21kW and 17kW without substrate temperature control. VPS101 was sprayed at 21kW with a large metal plate attached to substrate to maintain the substrate temperature.

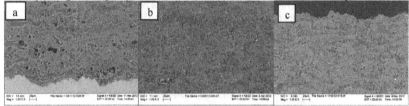

Figure 3: SEM of VPS samples: a) VPS003, b) VPS004, c) VPS101

### 3.3 Thermoelectric properties of Mg₂Si

Thermoelectric properties include *thermal conductivity, electrical conductivity* and *Seebeck coefficient*. The *power factor* is the combination of Seebeck and electrical conductivity given by $\sigma S^2$. Due to the low coating thickness of Mg$_2$Si samples prepared by thermal spray, not all of the samples could be characterized for the thermoelectric properties. We have shown reduction in thermal conductivity [15] before and the followings are some Seebeck results for APS and VPS samples.

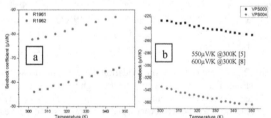

Figure 4: Seebeck coefficient of un-doped Mg₂Si by a) APS and b) VPS

Figure 4 shows the Seebeck coefficient of undoped $Mg_2Si$ prepared by APS and VPS. Typically, the higher purity thermoelectric gives the higher Seebeck coefficient. Our sample has MgO, Si and/or Mg in the coating, which may be responsible for the significant decrease in the Seebeck coefficient. The result shows that undoped $Mg_2Si$ is n type semiconductor and VPS has higher Seebeck which may due to higher purity.

## 4. CONCLUSIONS

$Mg_2Si$ layers have been synthesized by thermal spray, which has been shown to be an effective way to reduce the thermal conductivity of material in our previous investigation due to phonon scattering by pores and cracks in the coating. However, the air-sensitive element Mg can form large amount of oxide during the spray process, which may decrease Seebeck coefficient. VPS results showed an improvement in the reduction of oxidation thus increased the Seebeck coefficient. But high temperature due to poor heat transfer during the VPS process causes Mg to evaporate, which results in high concentration of Si in the coating. XRD results clearly showed that low temperature and high speed is useful to reduce the oxide in spray. VPS can achieve relative high Seebeck coefficient by choosing appropriate power.

**ACKNOWLEDGMENTS**

The authors gratefully acknowledge financial support for this work from NSF CBET #1048744 and NYSERDA under Contract 21180. The authors wish to thank Jim Quinn, Bo Zhang and Chao Nie from Stony Brook University for characterization of $Mg_2Si$.

**REFERENCES**
1. Morris, R.G., R.D. Redin, and G.C. Danielson, Phys. Rev.,109, (1958)
2. LaBotz, R.J., D.R. Mason, and D.F. O'Kane, J. Elec. Soc.,110, 127-134. (1963)
3. Zaitsev, V.K., Phys. Rev. B, 74(4), 045207. (2006)
4. Riffel, M. and J. Schilz., 15th Intl. Conf. (1996)
5. Jung, J.-Y. and I.-H. Kim, Elec. Mat. Lett., 6(4), 187-191(2010)
6. Kajikawa, Proc. ICT '97. XVI Intl. Conf. (1997)
7. Kajikawa, Proc. ICT 98. XVII Intl. Conf. (1998)
8. Tani, J.-I. and H. Kido, Phys. B: Cond. Mat.,364(1-4),218-224, (2005)
9. Akasaka, M., et al., J. Cry. Grow.,304(1), 196-201, (2007)
10. Tamura, D., et al., Thin Solid Films, 515(22, 8272-8276, (2007)
11. Q. Li, Z.W. Lin, and J. Zhou, J. Elec. Mat., 38(7), 1268-1272, (2009)
12. S. Sampath, JTST, 19(5), 921-949, (2010)
13. J. Zhou, Q. Jie, L.J. Wu, I. Dimitrov, Q. Li, and X. Shi, J. Mater. Res., 26(15), (2011)
14. J. Longtin, L. Zuo, et al. JTST (unpublished)
15. B. Zhang, L. Zuo, G. Fu, et al. Proceeding of Heat Trans. Conf. (2012).
16. Schilz, J., et al. J. Mat. Sci. Letter.,17(17), 1487-1490, (1998)
17. Ueno, K., et al. Proc. ICT 98. XVII Intl. Conf. (1998)
18. Takahashi, Y., et al. 18th Intl. Conf. (1999)
19. Cherigui, M., N.E. Fenineche, and C. Coddet, Surf. Coat. Tech.,.192(1), 19-26, (2005)

Mater. Res. Soc. Symp. Proc. Vol. 1490 © 2013 Materials Research Society
DOI: 10.1557/opl.2013.123

# Preparation and evaluation of the n-type PbTe based material properties for thermoelectric generators

Tse-Hsiao Li[1], Jenn-Dong Hwang[1], Hsu-Shen Chu[1], Chun-Mu Chen[1], Chia-Chan Hsu[1],
Chien-Neng Liao[2], Hsiu-Ying Chung[3], Tsai-Kun Huang[4], Jing-Yi Huang[4], Huey-Lin Hsieh[4]

[1] Material and Chemical Research Laboratories, Industrial Technology Research Institute, Hsinchu, Taiwan, ROC.
[2] Materials Science and Engineering, National Tsing Hua University, Hsinchu, Taiwan, ROC.
[3] Materials Science and Engineering, Feng chia University, Taichung, Taiwan, ROC.
[4] China Steel Corporation, Taiwan, ROC.

## ABSTRACT

Owing to energy conservation of waste heat, Lead telluride, PbTe, based materials have promising good thermoelectric properties around a range of middle temperature (Fig. 1, from 300 to 600°C), due to their high melting point, fine chemical stability, and the high figure of merit Z. The general physical properties and factors affecting the figure of merit have been reviewed. This research is focused on the n-type of PbTe materials and collocated with analysis of densities, hardness, elastic modulus, and thermoelectric properties thermoelectric figure of merit $ZT=GS^2T/\kappa$ (where G is electrical conductivity, S is Seebeck coefficient , T is absolute temperature, and $\kappa$ is thermal conductivity). Room temperature hardness and Young's modulus are measured by nano-indentation. In this study, the hot-press compacts under the pressure of 4 ton/cm$^2$ can reach the maximum density about 8.2 g/cm$^3$, and hardness and elastic modulus are 0.6 GPa and 70 GPa, respectively. The figure of merit value (ZT) of PbTe in low temperature (around 340°C) was found about 1 with carrier concentration above $10^{19}$ cm$^{-3}$. These results also indicate that the powder metallurgy parameters provide potentialities for further increase of the high efficiency of energy conversion in PbTe materials.

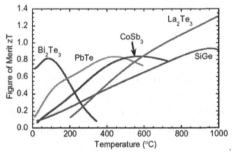

Fig.1 Performance of the established thermoelectric materials

**Introduction**

In general, thermoelectric effect is scientific and technological interest due to its widely application possibilities especially in power saving. Recent developments in theoretical studies on the thermoelectric effects, as well as the newly discovered thermoelectric materials provide new opportunities for wide applications[1-3]. There were great efforts to explore more effective materials, especially in semi-conductors. The theory identified at that time that the figure of merit for thermoelectric application is $ZT = TS^2G/\kappa$, where T is absolute temperature, S is the Seebeck coefficient, G is the electric conductivity, and $\kappa$ is the thermal conductivity. High ZT thermoelectric materials should be large S, high G, low $\kappa$, and could be found in semiconductor materials with carrier concentration of about $10^{19}$ cm$^{-3}$. One type of these materials, PbTe is based on the strongly correlated electron systems, and it also has ionic covalent bond with cubic sodium chloride-type lattice. The high ZT is also attributed to the doping, and introducing a strong enhancement of the density of states (DOS) due to a resonant state near the Fermi level that results in a significant enhancement of the Seebeck coefficient [4]. The thermoelectric properties, such as carrier concentration, Seebeck coefficient, mobility, electrical conductivity, and thermal conductivity are usually determined by the stoichiometry, such as: doping element concentration, and density, defects, etc. PbTe-based compounds include well-known n-type and p-type compounds for thermoelectric applications among the 200℃ to 600℃ temperature range.

Lead telluride, PbTe has a face centered cubic structure. At room temperature, PbTe is brittle material and also easily cleaved along the planes. Therefore, here would be still some efforts on changing different process conditions for the sake of making stronger devices. This study is also concentrated on the mechanical and thermoelectric properties of n-type PbTe samples which have a hole concentration above $1*10^{19}$ cm$^{-3}$. For determination of the free carrier concentration, Hall effect measurements were performed under normal pressure and temperature. The ZT value of n-type PbTe was found to be one of the most suitable thermoelectric materials until now under the usage on the middle-temperature ranges.

**Experimental**

2.1 Sample preparation

The n-type PbTe alloys were carried out by conventional powder metallurgy method, and synthesized by stoichiometric amounts of single elements Pb, Te with 99.99% high purity, then mixed for the desired compositions of Pb : Te = 1 : 1 (doped with less amount of high purity halogen) in the quartz tubes. According to Pb-Te phase diagram [5] indicates the presence of the

180

PbTe compound that is congruently melted around temperature 924°C. Therefore, the prepared admixtures were sealed in an evacuated high purity quartz tubes, and melted in high temperature (above 924°C). During the synthesis process, the sealed tubes were rocked continuously to ensure good homogeneity of melted ingots. The melted ingots in the tubes were then cooled to obtain good thermoelectric qualities. The solidified ingots were crushed, pulverized to powders, and hot pressed in a steel die at 380 °C under the pressure of 4 ton/cm$^2$.

2.2 Density test

The density of the thermoelectric samples was measured by Archimedes principle. The buoyant force on a submerged object is equal to the volumn of the object displaced in the water. Thus we can obtain the density of the hot-pressed sample from the equation as below.

2.3 Thermoelectric properties and Hall effect measurement

All of the measurements of thermoelectric property were performed on the self-assembled thermoelectric analysis instrument in Industrial Technology Research Institute (ITRI). The instrument is based on the principle of the figure of merit $ZT=GS^2T/\kappa$. The specimens were prepared in the shape of cuboid (8-8-12 mm$^3$).

The charge transport properties are measured using the Hall effect based on van der Pauw method [6], which electrical resistivity and carrier concentration、mobility are measured at the same time. The prepared hot-pressed specimen was cut into a sheet under 1mm thickness, and then utilizes four probes evenly spaced around the sample.

2.4 Hardness and Elastic modulus test

The hardness and elastic modulus values of all specimens were tested on MTS Nano Indenter XP Nano-indentation system and Nano-mechanical microscopy.

**Results and Discussion**

3.1 Density

All hot pressed PbTe samples can reach the density above 8.2 g/cm$^3$, but cold-pressed PbTe samples (with 2 hours sintering process) just attain 7.7 g/cm$^3$ (shown in Fig. 2). It also represents that the thermoelectric specimens with hot-pressing process exhibit higher density. However, the cold-pressed process may produce residual stress, then sintering procedure promotes the bonding of particles by application of heat, and also induces the voids and defects in the materials. In comparison with cold pressing process, the addition heat from hot pressing process can add to

181

the plasticity of the particles, and permit more intimate bonding and higher density simultaneous with high pressure.

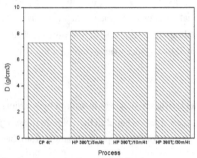

Fig.2 PbTe density dependence of different press conductions

### 3.2 Thermoelectric property and Hall effect measurement

Fig. 3-5 shows temperature dependences of S, G, κ and ZT of optimized thermoelectric material based on n-type of PbTe thermoelectric materials.

From the previous study, the figure of merit could be influenced by the hot-pressing conduction. However, the excess of pressing time would cause the increase of thermal conductivity. Due to the sintering process would increase the bond between the particles and strengthens a powder compact and also causes the increase of the grain size. On the other hand, it would also decrease the amount of the boundary with the excess of sintering time. Therefore, the increasing of the grain size would cause the raising of the thermal conductivity, and result in lower Z indirectly. As mentioned above, due to the high density of hot-pressed PbTe samples, the electrical conductivity can reach about 1500 1/ohm-cm, and lead the Seebeck coefficient -205 μV/K. Furthermore, the thermal conductivity could still stay below 3.55 W/ K-m. As a result, the measured ZT value shows that thermoelectric materials can attain ZTmax ≒ 1 around temperature range from 25 to 340℃. From Table 1, The Hall effect measurements can also reveal the charge transport properties.

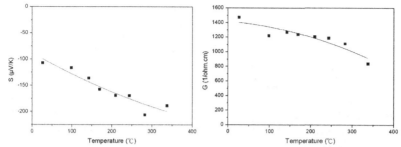

Fig.3 The Seebeck coefficient (S, left) and electrical conductivity (G, right) as a function of temperature (25-340℃) for n-type PbTe hot-pressed samples

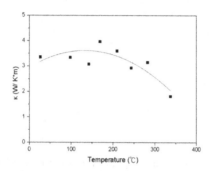

Fig.4 The thermal conductivity (κ) as a function of temperature (25-340℃) for n-type hot-pressed PbTe samples

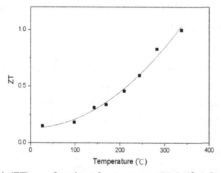

Fig.5 The figure of merit (ZT) as a function of temperature (25-340℃) for n-type hot-pressed PbTe samples

Table 1 Hall effect measurement

| Material | Carrier conc. (1/cm³) | mobility (cm²/V-s) | Resitivity (Ω-cm) |
|---|---|---|---|
| PbTe | -1.16E+19 | 2.21E+02 | 2.44E-03 |

3.3 Hardness and Elastic modulus

The hardness and elastic modulus measurement results of the hot-pressed PbTe specimen were 0.7 GPa and 64 GPa. (Fig. 6)

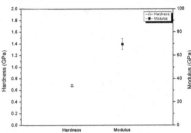

Fig.6 PbTe hardness and modulus tests

**Acknowledgments**

The authors would like to express sincere gratitude to Department of Industrial Technology (DoIT), Ministry of Economic Affairs (MOEA) of Taiwan; Pro. Chien-Neng Liao, Materials Science and Engineering, National Tsing Hua University; Tsai-Kun Huang, Jing-Yi Huang, Huey-Lin Hsieh, China Steel Corporation; Ass. Prof. Hsiu-Ying Chung, Materials Science and Engineering, Feng chia University for supporting this research.

**References**
1. CRC Handbook of Thermoelectrics, Edited by D.M.Rowe, CRC Press LLC., USA, 1995.
2. H. J. Goldsmid, Thermoelectric Refrigeration, Pion Ltd., London, 1986.
3. E. S. Toberer, A. F. May and G. J. Snyder, Chem. Mater., 2010, 22, 624–634.
4. J. Heremans, V. Jovovic, E. Toberer, A. Saramat, K. Kurosaki, A. Charoenphakdee and G. J.
5. Snyder, Science, 2008, 321, 554–557.
6. Bulletin of Alloy Phase Diagrams, Vol.10, No. 4, 1989

Mater. Res. Soc. Symp. Proc. Vol. 1490 © 2013 Materials Research Society
DOI: 10.1557/opl.2013.53

# Power generation performance of π-structure thermoelectric device using NaCo₂O₄ and Mg₂Si elements

Tomoyuki Nakamura[1], Kazuya Hatakeyama[1], Masahiro Minowa[1], Youhiko Mito[2], Koya Arai[3], Tsutomu Iida[3] and Keishi Nishio[3]
[1]SWCC SHOWA CABLE SYSTEMS CO., LTD. 4-1-1 Minami-Hashimoto Chuo-Ku Sagamihara-shi, Kanagawa-ken, 252-0253 Japan
[2]Showa KDE Co. Ltd. KDG AKASAKA bidg. 4F, 3-17-1, Akasaka, Minato-ku, Tokyo, 107-0052 Japan
[3]Department of Materials Science and Technology, Tokyo University of Science, 2641 Yamazaki, Noda-shi, Chiba, 287-8510 Japan

ABSTRACT

Thermoelectric power generation has been attracting attention as a technology for waste heat utilization in which thermal energy is directly converted into electric energy. It is well known that layered cobalt oxide compounds such as $NaCo_2O_4$ and $Ca_3Co_4O_9$ have high thermoelectric properties in $p$-$type$ oxide semiconductors. However, in most cases, the thermoelectric properties in $n$-$type$ oxide materials are not as high. Therefore, $n$-$type$ magnesium silicide ($Mg_2Si$) has been studied as an alternative due to its non-toxicity, environmental friendliness, lightweight property, and comparative abundance compared with other TE systems. In this study, we fabricated π-structure thermoelectric power generation devices using $p$-$type$ $NaCo_2O_4$ elements and $n$-$type$ $Mg_2Si$ elements. The $p$- and $n$-$type$ sintering bodies were fabricated by spark plasma sintering (SPS). To reduce the resistance at the interface between elements and electrodes, we processed the surface of the elements before fabricating the devices. The end face of a $Mg_2Si$ element was covered with Ni by SPS and that of a $NaCo_2O_4$ element was coated with Ag by silver paste and soldering.

The thermoelectric device consisted of 18 pairs of $p$-$type$ and $n$-$type$ legs connected with Ag electrodes. The cross-sectional and thickness dimensions of the $p$-$type$ elements were 3.0 mm × 5.0 mm × 7.6 mm (t) and those of the $n$-$type$ elements were 3.0 mm × 3.0 mm × 7.6 mm (t). The open circuit voltage was 1.9 V and the maximum output power was 1.4 W at a heat source temperature of 873 K and a cooling water temperature of 283 K in air.

INTRODUCTION

The worldwide depletion of energy resources is currently a serious problem, and it is extremely important that we reduce our energy consumption in order to save what little energy we have left. A related problem is that carbon dioxide generated when fossil fuel is burnt causes enormous amounts of greenhouse gases. A practical method of using clean energy with only a small load on the environment is needed in order to prevent global warming. In recent years, an attractive candidate for solving this energy problem has been thermoelectric power generation technology that can make effective use of unused energy. The thermoelectric generator is a technology that can convert thermal energy into electrical energy directly. It is an effective use of energy since we can convert waste heat energy, which is difficult to store, into electrical energy. Moreover, it is a clean method of generating electricity that does not emit $CO_2$ gas when it generates power.

So far, the materials with the highest thermoelectric performance have tended to be intermetallic compounds consisting of heavy metals such as Bi, Te, Sb, and Pb. However, there are few deposits of such heavy metal elements, and they are concerned about the hazard. For a while it seemed that it was necessary to take practical measures to ensure the steady supply and safe use of these materials. Therefore, in more recent years silicide materials and metal oxides have been attracting attention as an alternative due to their low environmental impact. Silicide is a promising material because there are abundant reserves of the raw materials and it has low toxicity. There have been recent reports of a high-performance Uni-Leg device that used $Mg_2Si$ materials [1-3]. There have been other reports of a conventional $\pi$-type device that can easily adapt to different temperatures and that is easy to build that used $NaCo_2O_4$ oxide and $Mg_2Si$ metal [4,5].

In this work, our objective was to fabricate an inexpensive $\pi$-type thermoelectric device that can deal with large temperature differences. We selected $NaCo_2O_4$ and $Mg_2Si$ as the *p-type* and *n-type* thermoelectric materials, respectively. We chose these materials because they are layered cobalt oxides that exhibit high thermoelectric properties at high temperatures [6-8]. In addition, oxides are robust against oxidation at high temperatures, and the synthesis of precursor powder is suitable for mass production because a soft chemical process can be applied. The soft chemical process is low-temperature and non-pressure, so its environmental impact is small. Most n-type oxide materials do not have properties as high as the *p-type* materials, so $Mg_2Si$, which is not an oxide material, was chosen as the *n-type*.

EXPERIMENTAL

$NaCo_2O_4$ was used for the *p-type* and Sb-doped $Mg_2Si$ was used for the *n-type*. The sintered body of both elements was obtained by spark plasma sintering (SPS). They were cut in element form. The cross-sectional and thickness dimensions of the *p-type* elements were 3.0 mm × 5.0 mm × 7.6 mm (t) and those of *n-type* elements were 3.0 mm × 3.0 mm × 7.6 mm (t). An alumina plate cut to dimensions of 33 mm × 27 mm × 0.65 mm (t) was used as the device substrate. An electrode pattern of the substrate was printed on the screen with Ag paste and printed the Ag paste was solidified by heating. Another top electrode was formed by Ag tape (0.2 mmt). 18-pairs, half-skeleton π-structure devices were then fabricated using these elements (Fig. 1). The ratio of the element to the substrate size was 39%.

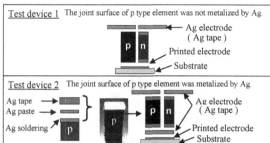

**Figure 1.** The fabricated device    Figure 2. Pattern diagrams of test device 1 and device 2

In the n-type elements, the joint surface was metalized by Ni using a monobloc SPS technique . In test device 1, we joined each element and upper electrode to the print electrode on the substrate. The $NaCo_2O_4$ element was joined by the Ag paste and the $Mg_2Si$ element was joined by a paste mixed with the Ag paste by low-temperature silver soldering. The elements were connected to the substrate at 873 K in a $N_2$ atmosphere (4% oxygen). In test device 2, we placed an Ag tape of the same size as test device 1 with the print electrode. The Ag tape electrode was formed on the substrate with the Ag paste. Silver soldering paste was applied to the joint surface of the $NaCo_2O_4$ element and the elements were connected to the Ag tapes with the Ag paste. These Ag tapes were connected at 1073 K for 2 h in a $N_2$ atmosphere (4% oxygen). The joint surface of the $NaCo_2O_4$ element was metalized by Ag (Fig 2). The elements were connected to the substrate under the same conditions as test device 1.

We used a ZEM-2 thermoelectric property measurement system (ULVAC-RIKO Inc., Japan) to determine the Seebeck effect and the electrical resistivity of each element from room temperature to 873 K in air. The power factors of the elements were then calculated from the measurement results. The internal resistance of the device at room temperature was measured by a general four-terminal method. The device was placed between a hot platen and a water-cooled heat sink and the I-V characteristics of the device were then measured (Fig. 3a). The alumina substrate side of the device was adjusted to the high-temperature side, and to insulate the Ag electrode and the heat sink electrically, silicone rubber was placed at the low-temperature side. The silicone rubber was 1.0 mm thick. Moreover after waiting to cool down to room temperature for 10 hours, the hot platen was again heated, and we also repeatedly measured the durability against heat cycle. The I-V characteristics and output of the device were measured by method of changing the load resistance of the circuit (Fig. 3b).

In the measurement of I-V characteristics and output of device, temperature of the heater ($T_h$) and the water flowing in the heat sink ($T_c$) were set to 873 K and 283K, and the temperature of each position was measured with a thermocouple thermometer. In the heat cycle test, preset temperature of heater and coolant water was same temperature condition as the above. All measurement was performed after holding in the state it heated for 1hour.

(a)            (b)

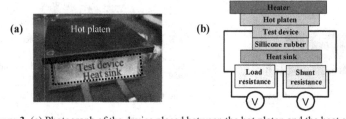

**Figure 3.** (a) Photograph of the device placed between the hot platen and the heat sink. (b) The I-V characteristic was measured while changing the load resistance with the circuit.

## RESULTS AND DISCUSSION

Figure 4 shows thermoelectric characteristic measurement results of each thermoelectric element. The power factor of NaCo$_2$O$_4$ was $0.75 \times 10^{-3}$ W/m· K$^2$ at 868 K, and the power factor

of Mg$_2$Si was $3.51 \times 10^{-3}$ W/m· K$^2$ at 577 K. In the case of the fabricated 18-pairs device, calculated values of open circuit voltage and internal resistance were estimated using values of

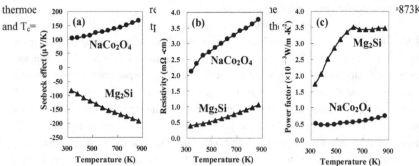

**Figure 4.** Thermoelectric properties: (a) Seebeck effect. (b) resistivity, and (c) power factor.
The internal resistance of the device at room temperature (RT) was 1.33 Ω for test device 1
and 0.42 Ω for test device 2 (Table 1). The resistance of the 18-pairs device calculated from the
resistivity of each element was 0.34 Ω. About the difference between the actual measured and
the calculated value of internal resistance of the device, test device 1 was large. This means that
connected resistance of the electrode and the $NaCo_2O_4$ element is large. In test device 2, it was
0.15 Ω, meaning the connection resistance had decreased, and as a result, the $NaCo_2O_4$ joint was
excellent compared with test device1.

**Table 1.** The internal resistance of the device at room temperature (RT)

| Device | Actual measured value of internal resistance at RT | Calculated value from resistivity of elements | Differences between measured and calculated value |
|---|---|---|---|
| Test device 1 | 1.33 Ω | | 0.99 Ω |
| Test device 2 | 0.42 Ω | 0.34 Ω | 0.08 Ω |

Figure 5 shows the device output measurement results of both test devices. The open-circuit
voltage of test device 1 was 1.75 V and that of test device 2 was 1.87 V. The maximum output of
test device 1 was 0.5 W and that of test device 2 was 1.4 W.

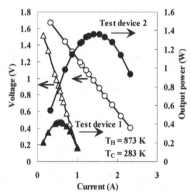

**Figure 5.** I-V characteristics and output power measurement for test devices 1 and 2.

**Figure 6.** Damaged $Mg_2Si$ elements after two heat cycles.

189

The open-circuit voltage calculated from the Seebeck effect of the element was 3.3 V and the value of the open-circuit voltage that had been obtained from the measurement was 57% of the output result. It seems the reason for this is that the temperature difference was small by the interfacial thermal resistance between the substrate and the hot platen and the thermal resistance of the alumina substrate and the silicone rubber. In test device 1, the $NaCo_2O_4$ elements separated easily after just one measurement because the bonding strength using only the Ag paste was so weak. The element did not separate so easily in test device 2, but after the heat cycle was repeated two times, cracks appeared in the $Ni/Mg_2Si$ element interface on the hot side. These cracks were probably caused by the different thermal expansion coefficients of the two elements. A photograph of the $Mg_2Si$ element damaged by the heat cycles is shown in Fig. 6.

## CONCLUSION

We attempted to fabricate a device that could be used at a higher temperature than BiTe devices. The device was comprised of $NaCo_2O_4$ and $Mg_2Si$ elements for low environmental impact. Experimental results showed that the connected resistance of the Ag-metalized $NaCo_2O_4$ elements decreased at a high temperature. A maximum output of 1.4 W was obtained at 873 K on the high-temperature side and 283 K on the low-temperature side. The output power density of this device corresponded to 1.5 $kW/m^2$. However, there is a problem with durability, as cracks eventually appeared in the $Mg_2Si$ element. Our future work will be improving the joint part durability of the device.

## REFERENCES

1. T. Sakamoto, T. Iida, Y. Taguchi, S. Kurosaki, Y. Hayatsu, K. Nishio, Y. Kogo and Y. Takanashi, *J. Electron. Mater*, 41, 1429-1435 (2012)
2. T. Nemoto, T. Iida, J. Sato, T. Sakamoto, T. Nakajima and Y. Takanashi, *J. Electron. Mater*, 41, 1312-1316 (2012)
3. T. Sakamoto, T. Iida, N. Fukushima, Y. Honda, M. Tada, Y. Taguchi, Y. Mito, H. Taguchi and Y. Takanashi, *Thin Solid Films*. 519, 8528-8531 (2011)
4. K. Arai, H. Akimoto, T. Kineri, T. Iida and K, Nishio, *Key Engineering Mater*, 485, 169-172 (2011)
5. K. Arai, M. Matsubara, Y. Sawada, T. Sakamoto, T. Kineri, Y. Kogo, T, Iida and K. Nishio, *J. Electron. Mater*, 41, 1771-1777 (2012)
6. I. Terasaki, Y. Sasago and K. Uchinokura, *Phys. Rev. B*, 56, 12685 (1997)
7. R. Funahashi, I. Matsubara, H. Ikuta, T. Takeuchi, U. Mizutani and S. Sodeoka, *Jpn. J. Appl. Phys*, 39, L1127 (2000)
8. M. Ito, T. Nagira, D. Furumoto, S. Katuyama and H. Nagai, *Scripta Materialia*, 48, 403-408 (2003)

Mater. Res. Soc. Symp. Proc. Vol. 1490 © 2013 Materials Research Society
DOI: 10.1557/opl.2013.318

# Design and implementation of a measurement system for automatically measurement of electrical parameters of thermoelectric generators

Dmitry Petrov, Fabian Assion and Ulrich Hilleringmann
Department of Sensor Technology, University of Paderborn, 33098 Paderborn, Germany

## ABSTRACT

The continues development of thermoelectric generators causes a permanent improvement of their characteristics. New types of thermoelectric generators can work at temperatures up to 1000 K. With this, special measurement equipment is needed to control the electrical parameters of the new developed specimens. The devices must be tested over the whole range of operating temperatures. For each temperature value a series of electrical measurements has to be performed. To establish the maximal output power of the thermoelectric generators, a load resistor with variable resistance has to be connected to the output of thermoelectric generator. The measurement system should measure the electrical current through the load resistor and the voltage over this resistor to determine the device parameters. A large amount of measurement data have to be collected and processed to evaluate the electrical characteristics of the specimen and to present them in graphical form, suitable for the comparison with others specimens.

## INTRODUCTION

The importance of regenerative energy systems has increased during the last years. The conversation between different energy forms, in electrical or combustion engines used in industry or transportation, will always cause some energy losses, mostly as thermal energy. Depending of the efficiency factor of the motor, this can be a considerable part of converted energy [1]. The efficiency of engines is theoretically limited by some fundamental laws, and practically by economical and technical reasons and cannot be improved much more. The other way to reduce the energy losses is to convert the dissipated power back into electrical energy by using thermoelectric generators.

In the past the upper limit of operating temperature of thermoelectric generators was limited to 150-200 °C [2,3]. The new generation of thermoelectric generators (TEG) with interconnects of titanium disilicide ($TiSi_2$) can operate at temperatures up to 1150 K [4]. To improve the efficiency of a TEG, it is essential to check their electrical parameters such as output current and voltage with different electrical load values over the whole range of operation temperatures. This allows to determine the maximal power point (MPP) of the TEG and thereby the maximal efficiency factor and the maximal electrical output power during the device operation.

The measurement equipment, used for the electrical tests of TEG, should allow measuring all above-mentioned electrical parameters over the specified range of operating temperatures.

## MEASUREMENT PRINCIPLE

Thermoelectric modules are formed by serial connection of p- and n- type semiconductors forming thermocouples, whereas the heat flow is in parallel through all thermocouples. The sides of TEG are called hot and cold side according to their temperature. Electrical power will be produced by the existence of a temperature difference between the hot and cold side of the TEG. The electrical power is used by connecting the load resistor to the end points of thermoelectric module (figure 1).

To measure the electrical parameters of TEG, the temperature difference must be known; it should be constant during the measurement process. This is assured by keeping the cold side on constant temperature by a water cooling system. The hot side of TEG is controlled by a precision temperature regulator, which is a part of the measurement system.

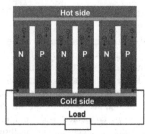

Figure 1: Basic connection of thermoelectric generator

The load is connected to the ends of TEG and has to be varied by the measurement system between the specified minimal and maximal value. On each load variation step the electrical current via the load and the electrical voltage over the load have to be measured. All there steps have to be done at well-defined temperatures [5]. Collected measurement data have to be stored in an internal memory until the end of the measurement process. Afterwards the data are transmitted to the host PC for further processing.

## REQUIREMENTS FOR THE MEASUREMENT SYSTEM

To cover the variety of output parameters of produced TEG specimens, the following requirements for the developed measurement system were established:

- The maximal measuring voltage of 4V (-4V to 4V up to the connection polarity)
- The maximal measuring current is 1A.
- The electrical load of TEG should be varied from 1 Ω to 160 Ω, in 1 Ω steps.
- Measuring accuracy is set to 1%.
- The temperature difference between the hot and cold sides of theTEG should be varied from 50 K to 500 K with an accuracy of 1 K.
- The measurement time for a specified temperature value should be less than 2 second.
- The complete measurement time over the whole temperature range should not be more than 30 minutes (depending on temperature and resistance measuring span).

## STRUCTURE OF THE MEASUREMENT SYSTEM

Figure 2 shows the structure of developed measurement system consisting of the following parts:

- **Water cooling system**, which keeps the temperature of the cold side constant (about 300 K, depending of settings);
- **Heating system** with precision temperature regulator, which allows to set up the temperature of the hot side in with high resolution and short reaction time;
- **Variable resistive load**, connected to the output of TEG consists of a chain of precision resistors, which are controlled by a control board;
- **Measurement system**, which is responsible for the measurement of the voltage and the current at specified load and temperature values;
- **Reference current source** for calibration of the measurement resistor chain, and for the detection of faults in switch elements;
- **Control board**, which coordinates the cooperative work of other system parts, collects measurement data and establishes the connection with the host PC;
- **Host PC software**, responsible for the measurement process parameter configuration and receiving the measured data.

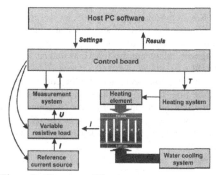

Figure 2: Structure of the measurement system

The hardware part of the measurement system is realized by a microcontroller board based on a PIC18 8-bit Microchip © microcontroller [6]. The built-in USB-interface of the microcontroller is used for the connection to a PC. The measuring voltage signal is converted into a digital form by the built-in 10-bit ADC. Other microcontroller periphery parts are used for communication to the other parts of the measurement system.

### MEASUREMENT PROCESS

The variable resistive load connected to the TEG output is based on a chain of precision resistors and switching elements as shown on figure 3. They allow a combination of serial and parallel electrical circuits. The serial connection of resistors insures the minimization of the required amount of resistors. But the resistance of the switching elements in closed position ($R_{ON}$) will also be added up for a small load and will falsify the defined load resistance.

The parallel connection of resistors in the range from 1 Ω to 4 Ω should minimize this problem by connecting only two serial switching elements with the load resistors (the second switch is used to bypass the parallel circuit). For the load values from 5 Ω, the joined resistance $R_{ON}$ of the switching elements is insignificant in comparison to the load resistance.

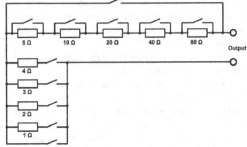

Figure 3: Circuit diagram of the variable resistive load

For the minimization of $R_{ON}$ of the switching elements, an electromagnetic relay is used in the resistive load circuit. This can assure an $R_{ON}$ of some mΩ at a maximal current up to 2 A [7]. These parameters are still unobtainable for todays semiconductor solid state switches. The disadvantages of electromagnetic relays are the longer switching time (about 3-5 ms) and the higher current consumption [8].

The oscillogram of switching the load, controlled by the control board is shown on figure 4. Channel 1 (below) shows the time for the switching operation reserved by the control board. Channel 2 (above) shows the voltage changes during the switching operation. The duration of switching a relay contact is about 2 ms, and the whole delay between two switching processes of the load element is about 6 ms. The part of this delay is used by the measurement system to measure the load voltage.

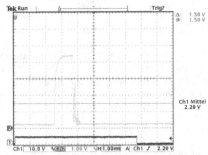

Figure 4: Oscillogram of the resistive load switching process

To minimize the measurement errors caused by electric and electromagnetic noises, some hardware and software methods are used in the measurement system. A low-pass filter with a cut off frequency of 1 kHz, implemented on the input of the analog-digital converter (ADC), removes short-term signal fluctuations. To minimize the internal ADC noise, a mean value filter with series of 10 measurements is implemented for each voltage measurement [9].

For the compensation of inaccuracies of the resistors in the resistive load circuit the reference current source is used. It allows to calibrate the real resistance value of each resistive element and uses this value for the calculation of the electrical current.

The whole measurement process for different load resistances at different temperatures is shown in figure 5. The measurement series for each temperature value takes about 1.6 second. The time to achieve the next temperature value takes some seconds, depending on the temperature differences between these values. The complete measurement for the whole temperature range will take between 20 and 30 minutes. The total time strongly depends on the specified measurement settings (temperature and resistance measurement span).

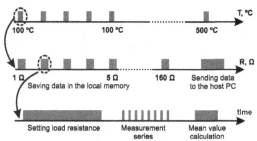

Figure 5: Series of measurements

## HOST PC SOFTWARE

The PC program is executed on the host computer and represents the graphical user interface (GUI) (figure 6) of the measurement system. It is responsible for the configuration of the measurement process and receives the measurement data from the hardware. A list of tasks, realized by the software is shown below:

- Establishing the data connection to the hardware of the measurement system, auto-detection of the connected hardware parts of the measurement system;
- Displaying the state of the measurement process and the actually measured values;
- Setting up the minimal and maximal values for the temperature and the resistive load resistor for automatic measurement;
- Setting up the temperature and resistance steps for automatic serial measurement;
- Starting and stopping the single measurement and the automatic measurement with specified parameters, showing progress of automatic measurement series;
- Allowing the user to save the measurement results in a text file.

The GUI is realized in C language with the usage of the standard WinAPI© (Windows Application Programming Interface) libraries. The communication with the host PC is performed by USB communication with Virtual Com-Port drivers.

Figure 6: Host PC control software

The measurement results from the saved text file can be imported to the electronic sheet by a special macro, programmed in Visual Basic language. This macro arranges the results in an electronic sheet according to the measurement temperature and load resistance.

CONCLUSION

The developed automatic measurement system for the electrical parameters of thermoelectric generators allows performing an accurate and fast measurement of new TEG specimens. The measurement process runs automatically for all specified temperatures and load resistor values. The implemented hardware and software solutions minimize the electrical noise during the measurement process and thereby provides accurate and reliable results in a temperature range up to 900 K. The software part of the measurement system and the data converting tool significantly reduce the time for collecting and converting measurement results for further analyses.

REFERENCES

1. Energetics, Energy Use, Loss, and Opportunities Analysis: U.S M&M; pp. 17; 2004.

2. R. Ahiska, S. Dislitas, Microcontroller Based Thermoelectric Generator Application. G.U. J. of Sci. 2006, 19, 131-146.

3. G. J. Snyder, Small Thermoelectric Generators. The Electrochemical Society Interface 2008, 54-56.

4. F. Assion, M. Schönhoff and U. Hilleringmann, "Titaniumdisilicide as High-Temperature Contact Material for Thermoelectric Generators", International Conference on Thermoelectrics, Aalborg, Denmark, 9th-12th July, 2012.

5. H. S. Han, Y. H. Kim, S. Y. Kim, S. Um, J. M.. Hyun, Performance Measurement and Analysis of a Thermoelectric Power Generator. IEEE 2010, 10, 1315-1325.

6. http://www.microchip.com/pagehandler/en-us/family/8bit/    8-bit PIC® Microcontrollers

7. http://www.maluska.de/service/hfd2.pdf / Signal relay, HFD2 Maluska Elektronik

8. http://www.clare.com/home/pdfs.nsf/www/an-145.pdf/$file/an-145.pdf,    Advantages of Solid-State Relays Over Electro-Mechanical Relays Application Note AN-145

9. D. Buoni, A. Senatore, S. Sannino, Analysis of the Frequency Response of the Mean Value Filter, International Conference on Engineering and Meta-Engineering: ICEME 2010

Mater. Res. Soc. Symp. Proc. Vol. 1490 © 2013 Materials Research Society
DOI: 10.1557/opl.2013.218

### Diffusion barriers for $CeFe_4Sb_{12}$/Cu thermoelectric devices

Laetitia Boulat, Romain Viennois, Didier Ravot, Nicole Fréty
Université Montpellier 2, Institut Charles Gerhardt, UMR 5253 CNRS-UM2-ENSCM-UM1,
cc 1504, Place E. Bataillon, 34095 Montpellier Cedex 5, France

ABSTRACT

The efficiency of a tantalum nitride interlayer as a diffusion barrier for $CeFe_4Sb_{12}$ thermoelectric material against electrode copper material has been investigated. The thermal stability of $CeFe_4Sb_{12}$/TaN/Cu stackings has been investigated after annealing at 600°C from a microstructural study. $CeFe_4Sb_{12}$ and Cu appear to chemically react through the formation of $CeCu_2$ and $Cu_2Sb$ phases whereas no reaction is observed for $CeFe_4Sb_{12}$ with TaN. This study showed that the TaN interlayer cannot inhibit the diffusion of Sb from the skutterudite substrate to the copper electrode but prevents the diffusion of Ce and consequently the formation of the $CeCu_2$ phase.

INTRODUCTION

The conversion of waste heat generated during industrial processes to electricity by use of thermoelectric devices is of particular interest [1-2]. Skutterudite compounds have been reported to be promising thermoelectric materials for applications in the [400°C-600°C] intermediate temperature range [3]. Doped or filled $CoSb_3$ and $CeFe_4Sb_{12}$-based skutterudites have been shown to be good candidates, respectively, for n and p-type legs of thermoelectric devices. However the performance of thermoelectric devices is strongly dependent on the joining of thermoelectric couples with metal electrodes as the conversion efficiency is greatly influenced by the contact resistance [4, 5]. A high electrical and thermal conductivity is required associated with a high interfacial mechanical strength [4]. Moreover the joining material has to be selected to avoid any interfacial reaction occurring during the device fabrication and use. The joining material has to play a role of diffusion barrier to limit the interfacial reactions which may be detrimental to the thermoelectric device performance.

The aim of this work is to study joining of $CeFe_4Sb_{12}$ material with Cu as electrode material. TaN is selected as the interlayer between $CeFe_4Sb_{12}$ and Cu due to its high thermal stability and electrical conductivity [6, 7] and due to its efficiency against Cu diffusion [8, 9]. This study focuses on the thermal stability of the $CeFe_4Sb_{12}$/TaN/Cu stacking after annealing at 600°C, which is reported to be the highest temperature of $CeFe_4Sb_{12}$ use for thermoelectric applications [3]. The efficiency of TaN as a diffusion barrier is determined from the evolution of the microstructure after annealing as a result of interfacial reactions. Interfacial reactions in $CeFe_4Sb_{12}$/Cu, $CeFe_4Sb_{12}$/TaN and $CeFe_4Sb_{12}$/TaN/Cu systems have been studied for this purpose.

EXPERIMENTAL DETAILS

$CeFe_4Sb_{12}$ skutterudite ingots were synthesized by direct reaction of stoichiometric amounts of Ce, Fe and Sb elements (Ce 99.9 % - Aldrich, Fe 99.9999 % - Alfa, Sb 99.9999 % - Alfa). The mixing was sealed in a carbon-coated silica tube under vacuum and heated to 1050°C over 48 h before being quenched in water. The samples were then straightly annealed at 680°C for 96 h [10, 11]. $CeFe_4Sb_{12}$ ingots were then cut into samples, 3 mm thick, to get substrates for TaN and Cu films to be deposited. These samples were polished (SiC 800) and cleaned with ethanol prior to deposition.

TaN films were deposited onto $CeFe_4Sb_{12}$ substrates by reactive radio-frequency (RF) sputtering under $Ar-N_2$ plasma using a Ta target (Johnson Matthey 2 in. diameter, 0.125 in. thick). The optimization of the process parameters to get stoichiometric TaN has been previously described [12]. TaN films, 1.5 μm thick, were deposited at room temperature with an Ar gas pressure of 6.9 Pa, a $N_2$ gas partial pressure of 2%, a sputtering power of 80 W and a deposit duration of 2 h. Cu layers were subsequently sputtered using a Cu target (Johnson Matthey 2 in. diameter, 0.125 in. thick) under Ar plasma. The Ar gas pressure, the sputtering power and the deposit time were 13 Pa, 90W and 1 h, respectively. Cu films, 1 μm thick, were deposited. The thermal stability of $CeFe_4Sb_{12}$/Cu, $CeFe_4Sb_{12}$/TaN and $CeFe_4Sb_{12}$/TaN/Cu samples has been studied after annealing at 600°C over 6 h under secondary vacuum.

The microstructure of the materials was observed using Scanning Electron Microscopy (SEM) (SEM S4500 - Hitachi) and the chemical composition was analyzed using energy dispersive X-ray spectrometry (EDS, Thermoelectron, Noran System SIX NSS 102). The crystalline structure was investigated through X-ray diffraction (XRD) using a θ–2θ diffractometer (Philips Xpert Pro MRD) with a Cu target X-ray tube working at 40 kV and 25 mA.

## DISCUSSION

The microstructural characterization of the $CeFe_4Sb_{12}$/TaN/Cu stacking in the as-deposited state was investigated prior to the study of the thermal stability.

### Microstructural characterization of as-deposited $CeFe_4Sb_{12}$/TaN/Cu stacking

A SEM cross-sectional view of the $CeFe_4Sb_{12}$/TaN/Cu stacking is reported in figure 1a. These thin films were found to exhibit a columnar growth perpendicular to the substrate surface which gives rise to a "cauliflower shaped" surface microstructure (figure 1b). This columnar growth is commonly observed in sputtered films considering the above process parameters, as predicted by the model of Movchan and Demchishin [13] and Thornton [14]. As a consequence, a columnar grain structure with inter-column voids and density-deficient grain boundaries is most likely to be intrinsic in sputtered films [15].

The XRD pattern of the raw $CeFe_4Sb_{12}$/TaN/Cu stacking is reported in figure 2a where the characteristic diffraction peaks of $CeFe_4Sb_{12}$, TaN and Cu phases are observed. The filled $CeFe_4Sb_{12}$ skutterudite crystallizes in the cubic structure of the Im-3 space group, where Fe atoms are located at the center of distorted octahedrals formed by Sb and Ce atoms [16]. The XRD pattern indicates that the material is made of the $CeFe_4Sb_{12}$ single phase without any secondary phase. The lattice parameter of $CeFe_4Sb_{12}$ was determined to be 9,135 Å from Rietveld refinement of a diffraction pattern performed on the substrate. This value is in good agreement with literature [11, 17]. The tantalum nitride thin film is observed to crystallize in the cubic δ-TaN phase with a NaCl-type structure, as previously reported [12, 18]. Copper films crystallize in the polycrystalline fcc-Cu phase with no preferred orientation.

198

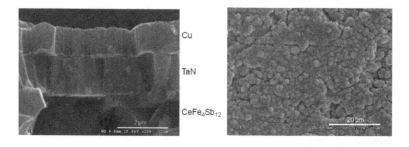

**a)**                                         **b)**

**Figure 1.** Cross-sectional SEM image of $CeFe_4Sb_{12}$/TaN/Cu stacking a) and surface SEM image of Cu thin film b)

## Thermal stability of $CeFe_4Sb_{12}$/TaN/Cu stacking

The efficiency of the TaN diffusion barrier was determined from the study of the thermal stability of the different stackings, i) $CeFe_4Sb_{12}$/Cu, ii) $CeFe_4Sb_{12}$/TaN, iii) $CeFe_4Sb_{12}$/TaN/Cu. These assemblies have been annealed under secondary vacuum at 600°C for 6 h and the evolution of the microstructure was observed in comparison with the as-deposited stackings previously described.

The study of the thermal stability $CeFe_4Sb_{12}$/Cu sample showed that $FeSb_2$, $CeCu_2$ and $Cu_2Sb$ phases are formed after annealing as evidenced from XRD analyses. The formation of $FeSb_2$ phase could be related to the diffusion of Sb to the substrate surface during annealing. The $Cu_2Sb$ and $CeCu_2$ phases are formed as a result of the chemical reactions between the substrate elements and the copper. These results point out that a diffusion barrier is needed to limit chemical reactions between the skutterudite substrate and the copper electrode. As previously described, tantalum nitride has been selected as an interlayer as it does not chemically react with copper up to 700°C [9, 19]. The thermal stability of $CeFe_4Sb_{12}$/TaN has then been investigated to study the chemical reactions after annealing at 600°C. Of particular interest, XRD analyses pointed out that TaN does not chemically react with $CeFe_4Sb_{12}$. Consequently TaN appears to be a potential material as a diffusion barrier between the $CeFe_4Sb_{12}$ substrate and the Cu electrode.

The barrier performance of tantalum nitride was investigated studying the thermal stability of the $CeFe_4Sb_{12}$/TaN/Cu stacking. Figure 2 shows the XRD pattern of the $CeFe_4Sb_{12}$/TaN/Cu sample in the as-deposited state and after annealing. As previously described, the $FeSb_2$ phase is observed after annealing due to the microstructure evolution of the skutterudite material. If the reactions of the skutterudite and copper are now considered, the $Cu_2Sb$ phase is formed as previously observed for the annealed $CeFe_4Sb_{12}$/Cu sample. SEM observation associated with EDS analyses confirm the formation of $Cu_2Sb$ precipitates which are observed at the surface of the sample (figure 3). The formation of the $Cu_2Sb$ phase may be related with the diffusion of Sb through the TaN interlayer and the subsequent reaction with Cu [20]. As previously reported, the inter-column voids associated to the TaN columnar growth may be assumed to be the main diffusion paths of Sb [9]. However, if $Cu_2Sb$ is formed when annealing $CeFe_4Sb_{12}/TaN/Cu$ samples, the XRD pattern shows that the TaN thin film allows to inhibit the formation of the $CeCu_2$ phase, in comparison with the annealed $CeFe_4Sb_{12}/Cu$ sample. Moreover it is worth being noted that the skutterudite phase is not observed anymore after annealing at 600°C in the XRD pattern (figure 2). This can be explained by the X-ray absorption from the reaction compounds formed, and more particularly from $Cu_2Sb$ precipitates of several micrometers thick (4 to 5 µm) observed at the surface of the sample. Indeed EDS analyses performed in the bulk of the substrate show the presence of the skutterudite phase after annealing.

These results point out that a TaN interlayer, about 1.5 µm thick, does inhibit Ce diffusion from the skutterudite substrate to the copper layer but does not appear to be efficient against Sb diffusion. Further investigations are needed to promote the TaN barrier efficiency against Sb diffusion, such as the effect of the diffusion barrier thickness or microstructure.

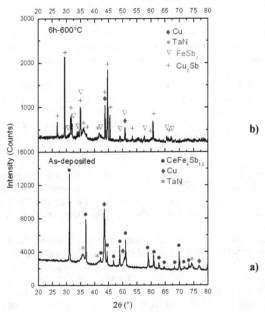

**Figure 2.** XRD patterns of as-deposited $CeFe_4Sb_{12}/TaN/Cu$ **a)** and annealed $CeFe_4Sb_{12}/TaN/Cu$ **b)**

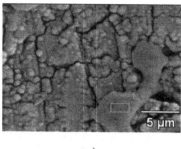

a)

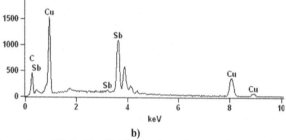

b)

**Figure 3.** SEM observation of CeFe$_4$Sb$_{12}$/TaN/Cu surface after annealing **a)** and EDS spectrum of Cu$_2$Sb phase (area 1) **b)**

## CONCLUSIONS

The thermal stability of CeFe$_4$Sb$_{12}$/TaN/Cu was studied after annealing at 600°C from a microstructural study. For a better understanding the interfacial reactions in the CeFe$_4$Sb$_{12}$/Cu and CeFe$_4$Sb$_{12}$/TaN annealed samples have been previously investigated. For CeFe$_4$Sb$_{12}$/Cu samples, XRD patterns evidenced the formation of FeSb$_2$, CeCu$_2$ and Cu$_2$Sb phases after annealing. The formation of FeSb$_2$ phase is related to the microstructural evolution of the skutterudite during annealing. The Cu$_2$Sb and CeCu$_2$ phases are formed as a result of the chemical reactions between the substrate elements and copper. The study of the CeFe$_4$Sb$_{12}$/TaN sample evidenced that CeFe$_4$Sb$_{12}$ does not chemically react with TaN, which shows the interest of TaN as a diffusion barrier.

The barrier efficiency of TaN was investigated from the microstructural study of annealed CeFe$_4$Sb$_{12}$/TaN/Cu samples. Whereas Cu$_2$Sb is formed after annealing, the formation of CeCu$_2$ is inhibited. Consequently a TaN thin film, 1.5 μm thick, appears to be efficient against Ce diffusion but does not inhibit the Sb diffusion and the subsequent formation of the Cu$_2$Sb phase. Further investigations are needed to study the effect of Cu$_2$Sb formation on the thermoelectric properties and to promote the TaN barrier efficiency against Sb diffusion varying the thickness and the microstructure of this barrier.

# REFERENCES

1. D.M. Rowe, *Renew. Energy* 16 (1999) 1251–55.
2. F.J. Disalvo, *Science* 285 (1999) 703–06.
3. P. F. Qiu, J. Yang, R. H. Liu, *Journal of Applied Physics* 109 (2011) 063713-1–063713-8.
4. K.T. Wojciechowski, R. Zybala, R. Mania, *Microelectronics Reliability* 51 (2011) 1198.
5. D. Zhao, H. Geng, L. Chen, *Int. J. Appl. Ceram. Technol.* (2011) 1–9.
6. D.K. Kim, H. Lee, D. Kim, *Journal of Crystal Growth* 283 (3-4) (2005) 404–8.
7. L. Liu, Y. Wang, H. Gong, *Journal of Applied Physics* 90(1) (2001) 416–20.
8. M. Stavrev, D. Fischer, C. Wenzel, *Thin Solid Film* 307 (1997) 79–88.
9. J. Nazon, B. Fraisse, J. Sarradin, *Applied Surface Science* 254 (2008) 5670–74.
10. L. Chapon, D. Ravot, J. C. Tedenac, *J. Alloys Compd.* 282 (1999) 58.
11. R. Viennois, L. Girard, D. Ravot *et al*, *Physical Review B* 80 (2009).
12. J. Nazon, J. Sarradin, V. Flaud, J.C. Tedenac, N. Fréty, *J. Alloys Compd.* 464 (2008) 526–531.
13. B.A. Movchan, A.V. Demchishin, *Phys. Met. Metallogr.* 28 (1969) 83–85.
14. J.A. Thornton, *Ann. Rev. Mater. Sci.* 7 (1977) 239–60.
15. S. Tsukimoto, M. Moriyama, M. Murakami, *Thin Solid Films* 460 (1/2) (2004) 222–26.
16. G.S. Nolas, D.T.Morelli, T.M. Tritt, *Annual Review of Materials Science* 29 (1999) 89.
17. W. Liu, Q. Jie, Q. Li, *Physica B* 406 (2011) 52-55.
18. C.S. Shin, D. Gall, Y.W. Kim, *Journal of applied Physics* 90 (6) (2001) 2879–85.
19. T. Laurila, K. Zeng, J.K. Kivilahti, *Microelectronic Engineering* 60 (2002) 71–80.
20. H.H. Saber, M.S. El-Genk, *Energy Conversion and Management* 48 (2007) 1383–1400.

**Thin-Films**

Mater. Res. Soc. Symp. Proc. Vol. 1490 © 2013 Materials Research Society
DOI: 10.1557/opl.2013.149

# Theoretical and experimental advances in $Bi_2Te_3$ / $Sb_2Te_3$ - based and related superlattice systems

M. Winkler[1], X. Liu[2], U. Schürmann[3], J. D. König[1], L. Kienle[3], W. Bensch[2], H. Böttner[1]

[1] Fraunhofer Institute for Physical Measurement Techniques IPM, Heidenhofstraße 8, D-79110 Freiburg, Germany
[2] Institute of Inorganic Chemistry, Christian-Albrechts-Universität zu Kiel, Max-Eyth-Str. 2, 24118 Kiel, Germany
[3] Synthesis and Real Structure, Institute for Materials Science, Christian-Albrechts-Universität zu Kiel, Kaiserstr. 2, 24143 Kiel, Germany

## ABSTRACT

Roughly a decade ago an outstanding thermoelectric figure of merit ZT of 2.4 was reported for nanostructured $Bi_2Te_3$/$Sb_2Te_3$-based thin film superlattice (SL) structures. The published results strongly fueled and renewed the interest in the development of efficient novel nanostructured thermoelectric materials. This review article shall give an overview over the most recent theoretical and experimental advances on $Bi_2Te_3$/$Sb_2Te_3$ SLs and related superlattice systems. The presented theoretical models are subdivided into electronic and phononic aspects. The experimental results are summarized with regard to the method used. A more detailed elaboration on structural and transport properties is given in the subsequent sections.

## INTRODUCTION

It is well known that roughly 60% of the energy resources are wasted as heat into the environment. A promising technology is the thermoelectric conversion of the waste heat into electricity. To obtain an efficient thermoelectric energy conversion a low thermal conductivity $\lambda$, a good electrical conductivity $\sigma$, and a large value for the Seebeck coefficient S are required. These material properties can be expressed by the thermoelectric figure of merit ZT which is defined as $ZT = \sigma \cdot S^2 T/\lambda$. ZT values > 1 are desirable for practical applications.

### A brief historical timeline for $Bi_2Te_3$ –based materials

The evolution of $Bi_2Te_3$ and related materials as thermoelectrics is displayed in Figure 1. The first well-documented systematic material screening investigating S and $\sigma$ of numerous compounds was carried out by Haken in 1910 [1]. Examining the systems Bi-Te and Sb-Te, he identified the phases $Bi_2Te_3$ and $Sb_2Te_3$ as promising thermoelectrics, starting their success story as room-temperature materials. Around the same time, Altenkirch gave a theoretical analysis of the problem of energy conversion using thermocouples in 1911 [2], showing that the performance of a thermocouple was affected by the involved materials' Seebeck coefficient, their electrical conductivity and their thermal conductivity, resulting in the equation for ZT. In the meantime, starting in the mid-40s the new material class of (thermoelectric) semiconductors came under research by different groups headed by such names as Schottky, Justi, Lautz, Goldsmid, Telkes and Ioffe [3,4,5,6,7]. In the 1950s, applied thermoelectric research in Europe, the United States and Russia by the pioneers Birkholz, Goldsmid and Ioffe progressed rapidly and first performance data of actual devices was presented. In 1954, Goldsmid et al. demonstrated a maximum temperature difference of 26 K with a Bi-$Bi_2Te_3$ thermocouple, followed by Birkholz, who managed to achieve a temperature difference of 38 K with $Bi_2Te_3$ in 1955 [3]. One way to improve this performance was proposed by Ioffe who was the first to introduce the concept of using solid solutions, i.e. alloys with isomorphous elements or compounds to decrease the ratio of thermal to electrical conductivity and thus improve ZT in 1956 [8,9,10]. Soon after, first researchers applied this concept to improve the performance of their devices, one of the first being Birkholz who increased the maximum achievable temperature difference to 70 K with the novel solid solutions $(Bi,Sb)_2Te_3$ and $Bi_2(Se,Te)_3$ [3]. Generally, applying proper contacts to thermoelectric materials is a key for the fabrication of efficient devices. Investigations carried out by Goldsmid revealed Ni to be an ideal contact material in contrast to Cu which tends to rapidly diffuse into $Bi_2Te_3$-based materials, drastically changing the electrical properties [4]. The rapid success in establishing device technology led to very optimistic assumptions for the potential of thermoelectric devices. In 1959, Zener proposed a maximum efficiency of 35 % as a, in his eyes, modest goal to strive for [11]. However, the enthusiasm began to wane when since the 1950-1960s no material could be found that actually exceeded a ZT of ~ 1. Note that while there is no theoretical limit on *ZT* [11], before the introduction of nanostructuring several authors have dealt with the limits of *ZT* under practical considerations. One of the early works estimates a practical maximum ZT of 1-2 [12]. Min and Rowe afterwards showed that due to restrictions in both and for all materials examined up to that date, ZT is limited to 2 at room temperature and ~ 4 at 1200 K [13]. They conclude that radically different concepts such as quantum well structures are required to increase ZT beyond that limit. The work of Vining in 1992 points in the same direction [11] – the urgent question why all materials discovered up to that point did not exceed a ZT of ~1 is posed and the author also proposes to examine new layered structures and superlattices.

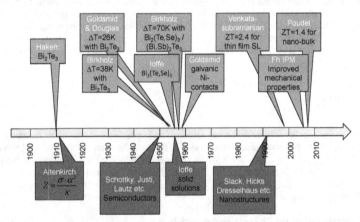

**Figure 1.** Timeline of the evolution of $Bi_2Te_3$ and related thermoelectric materials. Light grey = Performance milestones and important discoveries for $Bi_2Te_3$, Dark grey = Introduction of new general concepts.

Indeed and consequently in the 90s, research in the field of thermoelectrics upsurged because it was predicted by calculations that nanostructuring may significantly enhance the performance of thermoelectric materials [14]. The progress in the field of nanostructured thermoelectrics was reviewed in an abundance of different publications. An improvement was shown on bulk $Bi_2Te_3$-related materials. The fabrication of nanocomposites by ball-milling and subsequent hot pressing resulted in an increase of maximum ZT to 1.4. Even more impressive was the improvement of ZT at high temperatures: ZT at 250 °C was 0.8, which is a factor of ~4 improvement to a reference bulk crystal [15]. Similar experiments by our group showed that such material has superior mechanical properties compared to highly crystalline ingots [16].

Nevertheless, the best ZT values up to now have been achieved for thin films. For nanosized multilayers, especially Bi₂Te₃/Sb₂Te₃ superlattices (SL) for which ZT values of ~2.4 and ~1.5, respectively, were reported in 2001 [17]. Since this report a large number groups tried to reproduce or even to improve the properties of such SL systems. This review article is focused on theoretical and experimental results obtained during the last ~ 20 years in the field of Bi₂Te₃/Sb₂Te₃ nanosized layered systems.

## THEORETICAL RESULTS

### 1.) Effects of 2D nanostructuring on electronic properties

#### a) General considerations

Under electronic aspects, the basic intent behind nanostructuring is the modification of the dispersion relation and thus the density of states allowing new opportunities to vary S, σ, and λ quasi-independently for length scales small enough to give rise to quantum-confinement effects. Specifically for the electronic properties, the main goal for the introduction of nanostructures is the improvement of the power factor (PF) by changing the ratio of Seebeck coefficient to electrical conductivity in a favorable way. In this sense, Dresselhaus et al. presented their first calculations for ideal multi quantum wells (MQW) structures, e.g. infinite potential wells with separate wavefunctions and parabolic bands [14]. In the Boltzmann transport equation (BTE) based formalism, the modification of the density of states (DOS) into a staircase-like function induced by the nanostructuring lead to different expression for ZT. Not only could ZT be influenced by altering the Fermi energy (i.e. adjusting the doping level), but also by altering the width of the quantum wells. Assuming that the thermal conductivity can be calculated by setting the phonon mean free path equal to $d_A$ with current flow and arrangement of the ML parallel to the a-b plane and optimized Fermi energy, a factor of 3 increase over the ZT of bulk material (=0.5) was shown for small period lengths of ~ 1 nm (Figure 2). An experimental proof, refinement and extension of the concept discussed above was presented in [18] for PbTe/PbEuTe MQW structures. The ideal MQW structure was approximately realized since the barrier layers had much lower carrier mobility than the wells. Calculations taking into account multiple subbands together with experimental results proved a significant improvement of the ratio of Seebeck coefficient to carrier concentration and thus the product of S² with n and σ (PF). A *ZT* of 2.0 results from the PF improvement in the quantum wells.

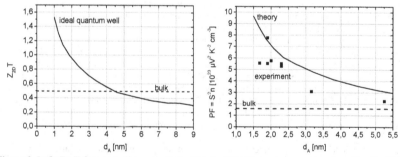

**Figure 2.** Left: $Z_{2D}T$ for a two-dimensional ideal Bi₂Te₃ MQW structure as a function of well width [14]. A significant improvement of ZT over the bulk value for small well widths is evident. Right: Significant improvement of the power factor compared to the bulk values for PbTe by PbTe/PbEuTe MQW structures [18].

However, the ideal MQW-based model has its limitations. Taking into account the transport properties of the whole system (wells *and* barriers) leads to a much lower effective ZT as discussed. In the above discussion for the PbTe/PbEuTe MQW, the thermal conductivity of the barriers was neglected – if duly taken into account, the $Z_{3D}T$ value will be much lower than the proposed $Z_{2D}T$ of 2.0 [18]. Furthermore, in a real SL structure, in contrast to the ideal MQW structure the barriers are not infinitely high. The well wave function penetrates into the barriers, where it decays exponentially and "leaks" into the other wells. A calculation of the subbands with a Kronig-Penney model [19] shows that for a decreasing well width the discrete energy levels broaden into minibands and the DOS is no more a sharp step function but "smears out", becoming more and more similar to a 3D dispersion relation. Thus, the improvement of ZT is smaller than calculated from the presented model for small well widths.

Koga et al. propose "carrier pocket engineering" as another quantum confinement mechanism to increase $Z_{3D}T$ for the whole GaAs/AlAs SLs [20]. Depending on the SL properties, different carrier pockets located at the X, L and $\Gamma$ points in the Brillouin zone (BZ) get occupied and contribute to carrier transport. In the optimum case of a 2 nm / 2 nm SL, the subbands at the X, L and $\Gamma$ points contribute to conduction and $Z_{3D}T$ is maximized, i.e. it is shown that $Z_{3D}T$ is highest in the case the carrier pockets at all three points contribute.

Another mechanism that may enhance ZT is electron filtering effects associated with ML type nanostructuring. Consider a bulk material with a DOS $\rho \sim E^{1/2}$. When the Fermi level moves deeply into the band, the differential conductivity becomes more and more symmetric with respect to the Fermi energy $E_F$ due to the flat square-root shape of the DOS [21]. A highly asymmetric differential conductivity in combination with high electron energies should yield a high Seebeck coefficient. This has first been realized by combining a metal with a wide-gap semiconductor to an SL, blocking low-energy electrons by the semiconductors bandgap [22]. The concept has been proven to work theoretically and experimentally on InGaAs / InGaAlAs – SLs, where an increase of the Seebeck coefficient by a factor of 2-3 was observed [23]. Electron filtering can also be achieved in bulk materials with nanoscale precipitates (filtering by change of scattering parameter [24]). Generally, all kinds of nanostructuring, i.e. 2D, 1D, 0D that induce sharp features of the DOS near the band edge can enhance the Seebeckcoefficient by increasing the asymmetry between hot and cold electron transport [21].

## b) Electronic properties of Bi$_2$Te$_3$ / Sb$_2$Te$_3$ - SLs

Recently, a series of results was published dealing with the electrical properties of Bi$_2$Te$_3$/Sb$_2$Te$_3$ stacks with special focus on the severe reduction of transport anisotropy in the SLs as proposed in [8] for cross-plane/in-plane electrical conductivity [17], a phenomenon that is still not well understood. In ref. [25], transport properties of Bi$_2$Te$_3$ and Sb$_2$Te$_3$ under strain were investigated via detailed first-principles electronic structure calculations based on density functional theory and the semi classical transport calculations based on the solution of the linearized Boltzmann transport equation (BTE). The method was validated by comparison with experimental data for the unstrained compound, yielding a very good agreement between theory and experiment. Subsequently, the anisotropy of both compounds was examined with respect to a lattice distortion caused by the strain within a layered heterostructure. In a further analysis [26], the anisotropy of Bi$_2$Te$_3$ and Sb$_2$Te$_3$ when stacked in a nanoscale heterostructure was examined. In the case of p-conduction, the heterostructure showed conductivity anisotropy *comparable to bulk material*. A clear preference for the in-plane transport direction was obtained.

The results are very interesting with regard to experimental results [17], where an elimination of the electrical conductivity anisotropy was found for SLs at distinct periods. No such effects could be confirmed by the mentioned theoretical calculations, Figure 3.
Recently, the ab initio calculations were refined and extended to a SL with 3 nm period consisting of different proportions of Bi$_2$Te$_3$ and Sb$_2$Te$_3$ [27]. In the frame of this work, quantum confinement effects

due to the band-gap differences of the two compounds were also taken into account. The authors still did not find a significant reduction of anisotropy compared to bulk materials. Thus, even assuming the most optimistic value (0.22 W/mK [17]) for the thermal cross-plane lattice conductivity, a maximum cross-plane ZT of only 0.9 was obtained at room temperature and 1.3 at elevated temperature, which is only roughly half the value reported in [17].

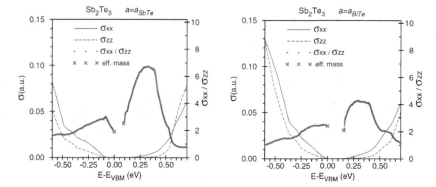

**Figure 3.** Left: Conductivity anisotropy of $Sb_2Te_3$ (thick line) without strain, corresponding to bulk. Right: $Sb_2Te_3$ under strain when stacked together with $Bi_2Te_3$, corresponding to the situation in a superlattice. Reprinted Fig. 5 with permission from [26]. Copyright (2011) by the American Physical Society.

## 2.) Effects of 2D nanostructuring on thermal properties

There are several models and approaches for modeling the thermal conductivity in SL systems and a considerable amount of publications has been devoted to this subject. The most significant models will be described here. Firstly, general applicable older models will be given and secondly, newer results specifically for $Bi_2Te_3/Sb_2Te_3$ SLs will be described.

### a) General considerations for SLs

Phonons traveling through a periodic SL structure undergo Bragg reflection if the condition $m\lambda_p = 2D\cos\theta$ is fulfilled [28]. Here, $\lambda_p$ is the phonon wavelength, $D = d_A + d_B$ the superlattice period and $\theta$ the angle of phonon incidence vector with the SL normal. The Bragg reflection produces band gaps or "stop bands" at the boundaries and centers of the Brillouin zone Figure 4. The width of these stop bands increases with decreasing period length and the phonon transmission rate (calculated for a finite SL with 15 periods) is strongly reduced at these frequencies. The depth and width of the transmission dips are correlated to those of the stop bands. For an increasing period number, the depth of the dips increases.

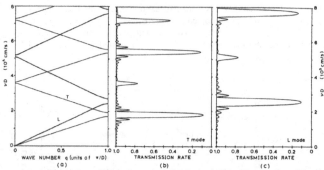

**Figure 4.** a) Dispersion relation for an infinite GaAs / AlAs SL with $d_A = d_B$ under normal phonon incidence. b+c) Transmission rate of transversal and longitudinal phonons. Transmission dips occur for frequencies that satisfy the Bragg reflection condition associated with the stop bands at $q = \pi/D$, the center and boundary of the folded Brillouin zone. Reprinted Fig. 3 with permission from [28]. Copyright (1988) by the American Physical Society.

Experimental proof of the phonon stop bands was carried out by Narayanamurti et al. [29] who determined the phonon transmission rate of a GaAs / GaAlAs SL. Dips in the transmission rates were found and their positions corresponded well to energies calculated under the assumption of Bragg reflection (0.93 meV). In analogy to optical dielectric photon filters, the authors coin the term of "dielectric phonon filters" for their SLs.

Chen et al. examined specular and diffusive scattering of phonons at interfaces based on a BTE-based calculation [30]. The calculations include an interface specularity parameter, $p = 0$ (1) for purely diffuse (specular) scattering, that allows to treat mixed specular and diffusive scattering. A critical parameter in the calculation of the thermal boundary resistance and thus thermal conductivity is the reflectivity and transmissivity of the layer-layer interfaces. For modeling these properties, the authors use the diffuse scattering limit or diffuse mismatch model (DMM), elastic acoustic mismatch model (AMM) and inelastic acoustic mismatch model. It is found that especially for SLs thinner than the phonon MFP, the effective $\lambda$ is to a large degree controlled by the interfaces and not the bulk scattering processes in the single layers. Thus the authors conclude that many good thermal conductors (such as Si or Ge) can be engineered to yield low-$\lambda$ structures. As expected, $\lambda_l$ is strongly dependent on the period thickness and also on the value of the specularity parameter. The more diffuse the boundary scattering, the lower the lattice thermal conductivity. By adapting p, in most cases good agreement to experimental data can be established.

Hyldgaard and Mahan evaluated Si/Ge phonon superlattice transport by (in contrast to previous works) taking into account not only modes that propagate either entirely within the SL plane or perpendicular to the interfaces [31]. In the latter case, the in-plane momentum $q = (q_x,q_y,0) = 0$. For $q = 0$, a simple one-dimensional chain model can be used while for $q \neq 0$ (phonons have an in-plane momentum) a three-dimensional model is necessary. It is instructive to compare the phonon frequencies that can propagate in the Si/Ge SL to those that can propagate in Ge or Si. Due to different force constants coupling the Si or Ge atoms (Si is harder), the phonon frequencies are generally higher in Si. It is found that for different in-plane momenta, there is a varying mismatch in phonon frequencies of Ge, Si and the SLs, leading to a phonon confinement and preventing their propagation.

Ren and Dow consider Umklapp-processes [32] that change the momentum of the phonon if the wave vectors of the colliding phonons result in a phonon with a wave vector outside of the first BZ. The size of the BZ is inversely proportional to the lattice constant, i.e. the size of the unit cell. Artificial SL structures can be seen as structures with very big unit cells and thus small BZs. This way, also phonons with smaller wave vectors can fulfil the conditions for a U-process. Thus, in addition to the bulk Umklapp-processes, in the SLs so called Mini-Umklapp processes can also take place, resulting in a lower thermal conductivity since in total, more phonons are affected.

## b) Considerations for $Bi_2Te_3$ / $Sb_2Te_3$ SLs

For his own MOCVD-grown $Bi_2Te_3$ / $Sb_2Te_3$ SLs, Venkatasubramanian explains the interconnection of thermal conductivity and SL period with models based on diffusive transport analysis and Bragg reflection at the interfaces [33]. Figure 5 shows the experimentally determined cross-plane lattice thermal conductivity. For a length scale, i.e. SL dimension < $l_{mfp}$, linear phonon heat conduction is assumed while for >$l_{mfp}$, the phonons would be scattered in another direction. It is concluded that a low-frequency cutoff $\omega_{cut} \propto 1/l_{mfp}$ can be defined for the SLs. The frequency cutoff increases for smaller $l_{mfp}$ and less phonons can participate in heat transfer, leading to lower thermal conductivity. The proposed near-complete (lossy) transmission of high-(long-) wavelength phonons is in agreement with previous observations on GaAs / AlGaAs SLs [29]. To explain the rise of thermal conductivity for very small period lengths it is proposed that the two layers get coupled when the cutoff wavelength approaches the thickness of both layers, making the effect of acoustic mismatch disappear. Thus, the long-wavelength phonons begin to be transported across the interface upon reducing the SL period length and λ rises again. A similar increase of λ for very small period lengths ≤ 4 nm was reported in [57].

Pattamatta and Madnia adopt the features of Chen's model that was described above to model the heat transfer in $Bi_2Te_3$ / $Sb_2Te_3$ SLs [34]. The BTE for the phonon intensity is solved with suitable boundary and interface treatment, i.e. the AMM and DMM model. Likewise, the scattering parameter p combines both models to calculate the total phonon intensity I from the AMM intensity $I_s$ and the DMM intensity $I_d$ by $I = p\, I_s + (1-p)\, I_d$. Results of calculations are given in comparison with experimental data from [33] in Figure 5. A p of 0.8 – 1 fits the experimental data appropriately. The model fails to predict the rise of thermal conductivity with small period lengths. The authors reason that this rise is due to the wave nature of phonons, see e.g. [33], while the presented model treats them as particles.

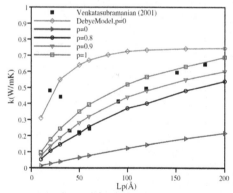

**Figure 5.** Variation of thermal conductivity k with period length Lp for $Bi_2Te_3$ / $Sb_2Te_3$ SLs. A satisfying agreement with experimental data is found for p = 0.8 – 1, but the rise in k for small period lengths is not predicted by the model. Reprinted from [34] with permission from Elsevier.

Takahashi et al. have used a matrix formalism to describe one-dimensional heat transport in multilayered films [35]. Experimental data from their films grown by PLD [57] serve for adjusting the model parameters. The authors apply their model to a SL considering two types of interfaces, i.e. sharp and disordered. In the sharp interface model, the interface resistance is $1.4 \times 10^{-8}$ m$^2$K/W. In the disordered interface model it is assumed that a thin layer of $(Bi_{0.5}Sb_{0.5})_2Te_3$ is sandwiched between the $Bi_2Te_3$ and $Sb_2Te_3$ layers. By modeling the SL with this interface region, the authors find a thickness of 1.4 nm which, interestingly, is higher than the thickness of the $Bi_2Te_3$ single layer in the high-ZT SL with 1 nm $Bi_2Te_3$ / 5 nm $Sb_2Te_3$ presented in [17].

## EXPERIMENTAL RESULTS

### 1. Growth techniques applied for the preparation of $Bi_2Te_3/Sb_2Te_3$ layered systems

A comprehensive overview about the different synthesis techniques to grow $Bi_2Te_3/Sb_2Te_3$ layers together with some details about the characterization techniques is compiled in Table 1. The structural and transport properties will be discussed in the following sections.

**Table I.** Deposition methods, substrates, deposition temperatures and measured transport properties for the preparation of $Bi_2Te_3/Sb_2Te_3$ layered systems. A feature shown in round brackets means results are only mentioned and not explicitly shown. If given in the same shade of grey the presented results are strongly related to each other and/or were obtained by the same group.
Abbreviations: superscript $^T$: support by theory; superscript $^c$: properties measured in cross-plane geometry; superscript $^D$: properties determined by device performance data. $ZT^H$: $ZT$ determined by the Harman method. no superscript: properties were determined from standard transport property measurements; RT = room temperature; AFM = atomic force microscopy, ATR-FTIR = attenuated total reflectance Fourier transform infrared spectroscopy, XRD = X-ray diffraction, HRXRD = high resolution XRD, T-XRD = temperature-dependent XRD, XRR = X-ray Reflectometry, RBS = Rutherford backscattering spectrometry, XPS = X-ray photoelectron spectroscopy, EL = ellipsometry, LEED = low energy electron diffraction, PS = phonon spectroscopy, RHEED = reflection high energy electron diffraction, EPMA = electron probe microanalysis, RS = Raman scattering, DF = device fabrication, c = element composition, gr-o = growth rate and optical constants, ts = thermal stability (systematic investigation), s = sound velocities, TEM = Transmission Electron Microscopy, SEM = Scanning Electron Microscopy; $D$ = device characterization and performance data; [c] $Bi_2Te_3$ – $Bi(Se,Te)_3$ superlattice system; [d] $Sb_2Te_3$ – $(Bi,Sb)_2Te_3$ superlattice system; [e] Electrochemical atomic layer epitaxy.

| Method | Substrate | Growth Temp. [°C] | Characterization techniques [a] | Transport properties [b] | Ref. |
|---|---|---|---|---|---|
| MOCVD | GaAs, (Al$_2$O$_3$) | - | (RBS,XPS, XRD, c) | $\mu$, S, PF, $\lambda^c$, ZT$^H$ | 36 |
| MOCVD | GaAs, Al$_2$O$_3$ | 350 | RBS,XPS, XRD, HRXRD, (c) | $\mu$, S | 37 |
| MOCVD | GaAs | 225 | SEM, TEM, (XRD) | $\mu$, n | 38 |
| MOCVD | GaAs | - | (TEM, XRD) | $\lambda^c$, $\lambda^T$, n | 33 |
| MOCVD | GaAs | - | - | $\lambda^c$,$\lambda^T$, $\mu$, $\sigma$, $\sigma^c$, S, ZT$^H$, D | 17 |
| MOCVD | GaAs | - | TEM | $\lambda^c$,$\lambda^T$ | 39 |
| MOCVD | GaAs | 350 | EL, gr-o, (XRD, LEED, XPS, RBS, c) | - | 40 |
| MOCVD | GaAs | - | DF, ts | ZT$^H$ | 41 |
| MOCVD | GaAs? | - | (EL) | ZT$^H$, D | 42 |
| MOCVD | GaAs? | - | - | (ZT$^H$), ZT$^D$, D, S$^D$ | 43 |
| MOCVD | GaAs | - | (XRD), PS | - | 44 |
| MOCVD | GaAs | - | PS, s, s$^T$ | ($\lambda$) | 45 |
| MBE epitaxial | Al$_2$O$_3$ | 340 | RHEED, XRD, AES | - | 46 |
| MBE epitaxial [c] | BaF$_2$ | Subref. | XRD, HRXRD, TEM, (SIMS) | $\lambda$, $\sigma$, S, ZT | 47 |
| MBE -Nanoalloy | BaF$_2$ and Si/SiO$_2$ | < 0°C  (~ RT) | SEM, XRD,TEM | $\mu$, n, $\sigma$, S, PF | 48 |
| MBE -Nanoalloy | BaF$_2$ and Si/SiO$_2$ | <0°C  (~ RT) | SEM, TXRD, TEM, SIMS, c | S,n | 49 |
| MER method | Si | RT | EPMA, XRD, XRR, SIMS, c | - | 50 |
| Sputtering | Si/SiO2/Cu | RT ? | AFM, RBS, RS, c | $\lambda^c$, $\sigma^c$, S$^c$, ZT$^c$ | 51 |
| Sputtering | Si/SiO2 | RT ? | XRD, XRR | $\sigma$, S, D | 52 |
| Sputtering –Nanoalloy [d] | Si/SiO2 | RT | SIMS, XRD, SEM, c | $\mu$, n, $\sigma$, S, PF,$\lambda^c$, ZT | 53 |
| Electro-chemistry | Al template | RT | - | - | 54 |

| EC-ALE [d] | Au | RT | XRD, SEM, ATR-FTIR, RS, c | - | 55 |
| Electro-chemistry | Si/Ti/Au | RT | XRD, SEM, TEM, c | - | 56 |
| PLD | Si | 250-400 | XRD, SEM, c | $\lambda^{c,}$ | 57 |

The first report about the outstanding performance of a $Bi_2Te_3/Sb_2Te_3$ SL system [2] may be regarded as an archetypical example of the "electron crystal – phonon glass" (PGEC) concept originally proposed by Slack [58]. According to the results the material exhibited simultaneously a large value for $\sigma$ across the stacking direction of the SL and a drastically reduced value for the thermal conductivity $\lambda$ in the same direction yielding the very high ZT values. This strong reduction is the main reason for the high figure of merit rather than electronic quantum confinement effects since such effects are only weakly expressed for the small conduction band offsets between the binaries.

Due to the big success in the application of MOCVD to fabricate the SLs, it is not a surprise the majority of articles dealing with $Bi_2Te_3/Sb_2Te_3$ SL systems are dedicated to this method (see Table I). Unfortunately, in ref. [2] results of the structural characterization with e.g. TEM or XRD were limited and therefore many questions remain unanswered to fully understand the structural and physical properties leading to the large ZT values. Hence data presented in [2] can only be partially compared with those of other research groups.

Another suitable method to prepare high quality $Bi_2Te_3/Sb_2Te_3$ SL systems is molecular beam epitaxy (MBE). Using this synthesis technique there is only one report where both structural and transport data were published for $Bi_2Te_3–Bi_2(Se,Te)_3$ SL on $BaF_2$ [47]. Intense works are underway at Fraunhofer IPM and the University of Kiel to fabricate epitaxial $Bi_2Te_3/Sb_2Te_3$ SLs using the MBE technique. It was also shown that an MBE setup can be used to prepare $Bi_2Te_3$-related binary thin films and SLs with the so-called nanoalloying method [48, 49, 53], details are given below. The nanoalloying method was also used successfully for sputtering of so-called "soft superlattice" structures ($Sb_2Te_3/(Bi,Sb)_2Te_3$ and $Bi_2Te_3/(Bi,Sb)_2Te_3$) exhibiting larger element start layer thicknesses [53]. In this case of p-conducting films the quality and definition of the nanostructure were much better than for films obtained by deposition with the MBE setup. The films showed a very strong c-axis texture, large values for PF and low thermal conductivities.

Besides the nanoalloying method, also co-sputtering has been applied to generate $Bi_2Te_3/Sb_2Te_3$ SLs [51, 52]. This deposition technique uses compound sputtering targets to deposit the SL instead of individual elements. The XRD data of such films indicate diffuse layers with partially rough interfaces. Interestingly, a quite high figure of merit ($\sim 0.8$) was reported for $Bi_2Te_3/Sb_2Te_3$ SLs irradiated with Si ions, mainly because the Seebeck coefficient became remarkably large after ion irradiation [51].

Another approach to synthesize $Bi_2Te_3/Sb_2Te_3$ and related SLs utilizes the electrochemical deposition method. The advantage of this method is that no high-vacuum setup is required and hence offers a cheaper way to fabricate such structures. The deposition of alternating layers of $Bi_{2-x}Sb_xTe_3$ (average x = 0.2) and $Sb_{2-y}Bi_yTe$ (average y = 0.7) with thickness ranging from 1 to 5 nm was done in a single bath setup. For the electrochemically deposited films, only few details concerning structural and electric properties were reported [54, 55, 56]. $Bi_2Te_3/Sb_2Te_3$ SLs were prepared by electrochemical atomic layer epitaxial growth, showing non-temperature stable superlattice satellites in the XRD [55]. Because no electrical data were reported for electrochemically fabricated SLs it cannot be judged whether this preparation route is suitable to prepare high-quality SLs exhibiting promising thermoelectric properties.

SLs of $Bi_2Te_3/Sb_2Te_3$ were also fabricated using PLD [57]. Using this technique very thin single layer thicknesses of $\sim 2$ nm were achieved leading to an extremely low thermal conductivity and an estimated ZT of 1.5 on the basis of some assumptions for the electrical properties.

A very interesting fabrication approach that does not yield superlattices in a classical sense (i.e. chemically modulated multilayers) was presented very recently. Atomically thin films of crystalline $Bi_2Te_3$ were prepared by the exfoliation method [59] exhibiting low thermal conductivity and high electrical conductivity leading to promising thermoelectric properties. Using the exfoliated thin flakes, "pseudosuperlattices" with a maximal thickness of $\sim 0.5$ mm were also prepared [60]. Measurements of the thermal conductivity demonstrated a reduction of the in-plane/cross-plane thermal conductivity by a factor of 2.4/3.5 compared to bulk $Bi_2Te_3$. The samples showed high Seebeck coefficients between 230 and 250 µV/K and also a low electrical resistance ($\approx 10^{-4}$ Ω m), suggesting promising thermoelectric properties.

## 2. Structural properties

Both compounds $Bi_2Te_3$ and $Sb_2Te_3$ crystallize in space group $R\bar{3}m$ with 15 atoms per unit cell according to the hexagonal setting. The atoms are located on two different sites (Bi and Sb: Wykoff site $6c$ with z = 0.40; Te on $3a$ and $6c$, z = 0.21) [61,62]. In stoichiometric samples all positions are fully occupied. The lattice parameters reported for $Bi_2Te_3$ vary slightly and for the a-axis the values are in the range 4.385 - 4.395 Å and between 30.44 and 30.5 Å for the c-axis. The analogous values for $Sb_2Te_3$ scatter around 4.264 Å for a and 30.46 Å for c. In ideal $Bi_2Te_3$-$Sb_2Te_3$ SLs individual $Bi_2Te_3$ and $Sb_2Te_3$ layers are stacked onto each other being characterized by a layered structural motif along the growth direction, i.e. the c-axis (Fig. 2). A so-called quintet or quintuple contains the sequence Te(2)-Bi/Sb-Te(1)-Bi/Sb-Te(2). The Bi/Sb-Te bonds are of ionic-covalent type and the Te(2)–Te(2) interaction between the layers are of van der Waals type [63]. The periodical van der Waals bonding interactions in the materials along the c-axis is a promising prerequisite for the deposition of high quality superlattice interfaces [37]. The layered structure also makes the material properties highly anisotropic.

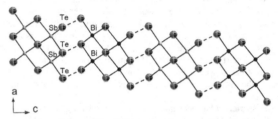

**Figure 6.** Idealized scheme for the atomic stacking sequence within a $Bi_2Te_3$/$Sb_2Te_3$ superlattice. The big, white and black solid circles represent Te, Sb and Bi atoms and the dashed lines indicate the van der Waals bonds.

According to XRD data of the materials prepared by the above mentioned methods most samples contain SL structures of pure $Bi_2Te_3$ and $Sb_2Te_3$, respectively while some authors observed slight deviations from the stoichiometric ratio.

### a) Structural characteristics of $Bi_2Te_3$/$Sb_2Te_3$ SLs prepared with MOCVD

Astonishing structural results have been reported for the SLs deposited on (100) GaAs. It is well known that the $V_2VI_3$ compounds with threefold symmetry can be grown on cubic (111) surfaces that also have threefold symmetry. There are several reports of successful epitaxial growth of $Bi_2Te_3$-based materials on cubic substrates such as $BaF_2$ and CdTe [64]. The atoms in the (111) plane can be considered as hexagonally arranged [65] and the cubic and hexagonal lattice constants $a_c$ and $a_h$ are related by $a_h = a_c/\sqrt{2}$. Under these considerations, e.g. (111) $BaF_2$ ($a_c = 6.196$, $a_h = 4.381$ Å) [64] shows a very good lattice match to $Bi_2Te_3$ ($a_h = 4.386$ Å) resulting in films with high structural quality [47,65]. However, these considerations do not apply to cubic (100) surfaces with fourfold symmetry such as GaAs. The problem of symmetry mismatch was not adressed specifically in the associated publication(s) where instead a lattice mismatch of 22 % was stated [38] that was apparently calculated from the lattice constants $a_h$ of $Bi_2Te_3$ and $a_c = 5.653$ Å of the GaAs substrate.
Despite these obstacles, the authors report to have obtained "single crystalline" films though it appears that rotational disorder was not excluded. Even high-quality epitaxial films grown by other groups have

been shown to display a slight degree of columnar twinning [65,66,67]. The authors prove the high quality of their films by a Fourier transform of the HRTEM image on the $Bi_2Te_3/Sb_2Te_3$ SL sample deposited on 2-3° misoriented (100) GaAs substrate and an additional XRD measurement. The XRD results presented are restricted to a $Bi_2Te_3$ film deposited on $Al_2O_3$ and the $2\theta$ range of 39° to 47°, where only the (00.15) reflection of $Bi_2Te_3$ is observed [37]. Due to the high crystalline quality reported, the samples produced by MOCVD exhibit a low degree of interdiffusion between $Bi_2Te_3$ and $Sb_2Te_3$ layers. TEM micrographs show clearly that the SL structure is well expressed, which is also confirmed from the presence of the satellite peaks in the double-crystal X-ray diffraction pattern [38]. In-situ spectroscopic ellipsometry carried out during growth confirmed "perfect superlattice growth with an abrupt interface between the two constituent films" [40]. The purported high quality of the SL structure is the predominant reason for the significant reduction of the thermal conductivity in the cross-plane direction. The lowest thermal conductivity was achieved when the dimension approaches 3 nm. Results concerning the details of the SL structure including nanoprobe chemical analyses were not presented.

## b) Structural characteristics of $Bi_2Te_3/Sb_2Te_3$ SLs prepared with MBE

In-situ RHEED observation of a $Bi_2Te_3/Sb_2Te_3$ SL deposited on sapphire (00.1) substrates which were kept at 340 °C were reported [46]. The RHEED patterns confirm epitaxial growth and indicate that the initial growth mode is three-dimensional, changing to a two-dimensional growth within minutes. In the powder pattern the $(00l)$ reflections are dominant due to a high level of $c$-axis orientation of the crystallites in the SLs. The high quality of the hetero-interfaces leads to satellite peaks in the diffraction pattern.

Homogeneous epitaxial SL films with thicknesses from 40 to 4000 Å were deposited on $BaF_2$ (111) substrates and the properties of the films were thoroughly characterized [47]. It was demonstrated that freshly cleaved $BaF_2$ (111) is a suitable substrate for epitaxial growth of $Bi_2Te_3$ films due to the small lattice mismatch between the $BaF_2$ (111) and $Bi_2Te_3$ basal plane. In the XRD pattern of a $Bi_2Te_3$ film only (00l) reflections of $Bi_2Te_3$ and (111) reflection of $BaF_2$ could be observed being a clear indication for a nearly perfect $c$-texture. In a further detailed study, the SL interfaces, the structure and also dislocations in the films as well as their relation to the SL properties were analyzed [68]. The structural analysis confirms the high quality of the films. Interestingly, a structural nanoscale modulation was detected which was named "natural nanostructure (nns)". The results indicate that the micro- and nanostructure, governed by the interplay of the SL structure, the nns and the dislocations directly affects the lattice thermal conductivity. Very recently, it was revealed that the nns can be switched on and off by argon ion etching [69]. A detailed study of the nns was performed with stereomicroscopy in the TEM and by image simulation [70]. The results indicate that the nns is a sinusoidal displacement field with a displacement exhibiting an amplitude of ca. 10 pm with a wave vector parallel to $\{10.10\}$ and corresponding wavelength of 10 nm. Video sequences recorded in a TEM demonstrate the behavior as well as mobility of dislocations in bulk $Bi_2Te_3$ [71].

Another method that is carried out with an MBE setup is nanoalloying. Nanoalloyed $Bi_2Te_3/Sb_2Te_3$ SLs were examined, revealing that the SL structure can be well formed in the as-grown state that is only partially crystalline [48]. A complementary set of electron microscopy techniques including analytical methods and Z-contrast imaging prove the chemical segregation of consecutive layers into $Bi_2Te_3$ and $Sb_2Te_3$, respectively. The period length was determined to 18 nm. The XRD data on the annealed samples show that only the desired $V_2VI_3$ phases were present after the annealing procedure and that an increase in annealing temperature (up to 250 °C) leads to larger grain sizes. Unfortunately, the desired strong $c$-texture was not observed for these samples by X-ray and electron diffraction. According to ex-situ (in furnace) and in-situ (in TEM) heating experiments [72], significant interdiffusion between the $Bi_2Te_3$ and $Sb_2Te_3$ layers was apparent through the annealing process.

215

## c) Structural characteristics of Bi$_2$Te$_3$/Sb$_2$Te$_3$ SLs prepared with Sputtering

The magnetron sputtering technique was used for the fabrication of Bi$_2$Te$_3$/Sb$_2$Te$_3$ multilayers using Bi$_2$Te$_3$ and Sb$_2$Te$_3$ targets as sources [52]. In the XRR curves, oscillations appeared at scattering angles indicating thicker layers than expected (10 nm). XRD data confirm that the 31 layer sample consists of the desired crystalline phase. Nevertheless, the level of c-axis texturing is low with about 1/3 of the crystallites contributing to the c-texturing and the remaining grains being oriented with the (10.10) s.o. axes along the surface normal.

Nanoalloyed superlattices can be fabricated using sputtering techniques. Our current research group had reported the investigation on the p-type Sb$_2$Te$_3$ / (Bi$_{0.2}$Sb$_{0.8}$)$_2$Te$_3$ SL, where each compound layer thickness was 25 / 12.5 / 6.25 nm, yielding a period length of 50 / 25 / 12.5 nm, respectively [53]. Layers of elements were sputtered onto Si/SiO$_2$ (100nm) substrates and then annealed to yield the desired phase and SL pattern. Different annealing temperatures ranging from 150 to 350 °C were applied. The XRD patterns show that the as-grown sample was amorphous and crystallized upon annealing. Rietveld refinements were carried out on the XRD data. The average crystallite size was calculated from the refined peak profile, showing an increment with increasing annealing temperature. Unlike reported in the MBE nanoalloyed SL in this system [48], the annealing process, judging from the XRD data, has evidently introduced the preferential growth of crystallites that are oriented with the c-axis perpendicular to the substrate. This preferred orientation became more pronounced at higher annealing temperatures. The degree of orientation could also be strongly increased by reducing the period length. For the smallest period length of 12.5 nm, even at a low annealing temperature of 150°C the degree of c-orientation was nearly absolute, comparable to epitaxial films. Research on the thermal conductivity of such low-periodic nanoalloyed structures will be published in the near future.

## d) Structural characteristics of Bi$_2$Te$_3$/Sb$_2$Te$_3$ SLs prepared with Electrochemical Deposition

The growth mechanism was studied on a SL sample with the average composition Bi$_{0.25}$Sb$_{0.16}$Te$_{0.58}$ deposited on gold substrate using the EC-ALE method [55]. It was suggested that the film is conformal with the substrate during initial growth and the platelet morphology develops during a later stage of film growth. In the XRD pattern the occurrence of satellite peaks (period length of 23 nm) is indicative for a high quality of the SLs. Applying Raman spectroscopy the authors were able to demonstrated a nearly perfect interface BiTe/SbTe bonding in the as-deposited films [55]. Nevertheless, the SL structure of individual alternating Sb$_2$Te$_3$/Bi$_2$Te$_3$ layers was destroyed during an annealing step due to interdiffusion of the elements leading to the formation of the ternary compound. Large scale Bi$_2$Te$_3$/Bi$_{2-x}$Sb$_x$Te$_3$ multilayers produced by optimized pulsed electrodeposition showed a periodic modulation of the Sb content [56]. Using Z-contrast imaging in STEM the multilayer structure shows a periodicity in the range of 10-30 nm. EDX mapping in scanning mode of the TEM allowed visualization of the multilayer morphology. Electron diffraction on the polycrystalline materials proves that the {01.5} planes of the multilayers are preferably oriented normal to the substrate plane.

## 3. Transport properties

### a) Transport properties of films grown by MOCVD

The transport data for a short-periodic (period length ≤ 10 nm) Bi$_2$Te$_3$/Sb$_2$Te$_3$ SLs was already reported in 1996, looking extremely promising [36] already at this early stage and strongly suggesting that MOCVD is a suitable method for the fabrication of such SLs. So-called asymmetric SL structures with variable period length and fixed Bi$_2$Te$_3$ thickness (15 Å) showed very high in-plane carrier mobilities of more than 600 cm$^2$/Vs together with large values for the in-plane Seebeck coefficient of ~ 200 – 270 µV/K, yielding an outstanding power factor of 59 µW/cmK$^2$. This large power factor exceeds that of state of the art p-type bulk material of ~40 $\mu$W/cmK$^2$. One should note that the electrical properties were achieved while simultaneously the cross-plane thermal conductivity (minimum value of 0.22 W/mK) was a factor of 4-7 smaller than that of bulk material. Additional phonon-scattering occurring in the periodic structure was announced to be responsible for the lower thermal conductivity compared to bulk material (see the theoretical considerations presented above).

Measurements of the carrier mobility and Seebeck coefficient of SLs [37] (thickness of the single binary layers not clearly mentioned) show slightly lower values than for the asymmetrical SLs mentioned above. According to the transport properties, MOCVD generates n-type conducting binary $Bi_2Te_3$ while binary $Sb_2Te_3$ is always p-conducting. Stacking p-type and n-type conducting materials onto each other compensation effects can be expected and were indeed observed on nanoalloyed SLs fabricated by MBE [48,49,73]. Theoretically, the Seebeck coefficients $S_A$ and $S_B$ of materials A and B with opposite signs cancel out each other according to [65]

$$S_{total} = (d_A\ \sigma_A\ S_A + d_B\ \sigma_B\ S_B) / (d_A\ \sigma_A + d_B\ \sigma_B) \quad (eq.1)$$

with $d$ as single layer thickness. Interestingly, in contrast to this expectation the Seebeck coefficients of the MOCVD grown SLs are very high and exceed 250 $\mu$V/K [36]. This observation is not fully understood and several possible explanation may be envisaged. It is possible that the interfaces of the SLs are not atomically abrupt and if the $Sb_2Te_3$ layers are thicker than the $Bi_2Te_3$ layers (asymmetric SLs), Sb may partially interdiffuse into $Bi_2Te_3$ yielding a ternary phase $(Bi_{1-x}Sb_x)_2Te_3$ which is p-conducting for x $\gtrsim$ 0.5 [74]. On the other hand interdiffusion is not necessary if the superlattice itself is asymmetric with very low dimensions like the stacks of 1 nm of $Bi_2Te_3$ on 5 nm of $Sb_2Te_3$ reported in [17]. Hence, the 1 nm thick layer of $Bi_2Te_3$ is surrounded by 2 nm of $Sb_2Te_3$, and a full 3 nm unit cell is generated consisting of

Te-Sb-Te-Sb-Te • Te-Bi-Te-Bi-Te • Te-Sb-Te-Sb-Te

(• = van der Waals type bonds). Such a SL may be viewed as a type of ordered alloy with composition $(Bi_{0.33}Sb_{0.67})_2Te_3$ which is p-type conducting. Hence, p-type conducting $(Bi_{0.33}Sb_{0.67})_2Te_3$ is stacked on p-$Sb_2Te_3$ and the compensation effect is eliminated yielding high Seebeck coefficients. Unfortunately, for a symmetric SL with identical layer thicknesses no Seebeck coefficients were reported [38] and one can only speculate that compensation effects may have led to small values for S. The in-plane carrier mobilities for the symmetric SLs are lower [38] than for asymmetric SLs presented in [36] but they are significantly larger than those reported for reference ternary alloys and range between that of the binaries. This observation suggests that charge carrier scattering at the interfaces of individual layers is significantly weaker than alloy scattering.

The electrical conductivity anisotropy in the SLs was studied in detail [17]. It is well documented that the bulk materials show a pronounced conductivity anisotropy. This anisotropy can be expressed as the ratio $\sigma_c/\sigma_{a-b}$ with $\sigma_c$ as conductivity parallel to the c-axis, corresponding to the cross-plane direction in a exclusively c-textured films and $\sigma_{a-b}$ perpendicular to the c-axis, corresponding to the in-plane direction in a film. This ratio is around 0.2-0.4 for bulk-$Bi_2Te_3$ and $Sb_2Te_3$ [63]. In contrast to this, the anisotropy ratio for $Bi_2Te_3/Sb_2Te_3$ SLs was reported to be around 1 or even higher. It was suggested that in SLs very small band offsets between $Bi_2Te_3$ and $Sb_2Te_3$ occurs leading to a weak confinement effect [17]. However, if this assumption is correct pure $Bi_2Te_3$ and $Sb_2Te_3$ should exhibit identical cross-plane and in-plane conductivity because the valence-band offsets between the quintuples forming the unit cell of these compounds are zero [75]. Theoretical calculations dealing with this phenomenon were presented above in the theory part.
The high $ZT$ values of 2.4/1.5 for p/n superlattices were determined with the Harman method for varying superlattice thicknesses. Difficulties in the measurements were noted by the author. A distinct voltage spike is observed in the measurements, complicating the determination of the ohmic part $V_0$ of

the Harman voltage. In the plots shown, for different thicknesses (Fig. 2b in [17]) some scattering of the data points is evident that may be associated with these complications.

High-performance thermoelectric coolers may be fabricated with the SLs because low contact resistances in the range of $10^{-8}$ $\Omega$ cm$^2$ were achieved [8]. Cooling power densities > 150 W/cm$^2$ [42] and maximum hot-cold side temperature differences $\Delta T$ of more than 55 K [43] or 70 K [17] were also demonstrated.

Wang et al. carried out measurements of phonon lifetimes by time-resolved pump-probe experiments. Both optical [44] and acoustic [45] phonons were examined. The oscillations of certain optical phonon modes can be directly made visible in the oscillatory change of reflectivity. On bulk materials, a very good agreement of the frequency of observed phonon modes with results from previous Raman scattering experiments was found. The coherent phonon vibration signal of the films shows an exponential decay whose time constant corresponds to the scattering rate. In comparison, the signal for the Bi$_2$Te$_3$ / Sb$_2$Te$_3$ SLs decays noticeably faster than that of the binaries, showing that the phonon lifetime in SLs is shorter and suggesting phonon-interface interactions. For Bi$_2$Te$_3$, Sb$_2$Te$_3$ and the SLs the determined scattering rates are 0.188, 0.295 and 0.357 THz, respectively, quantitatively confirming the lifetime reduction in SLs.

Likewise, acoustic phonon scattering was examined for long-wavelength phonons [45] by examining the reflectivity signals (Figure 7). The change in reflectivity is proportional to the phonon amplitude after traveling forth and back through the film with acoustic reflections at the film/substrate (binary) or film/buffer (SLs) interfaces. The authors found that upon increasing the film thickness, the phonon signal amplitude was hardly diminished for Bi$_2$Te$_3$ but reduced by 40-50 % for the SLs.

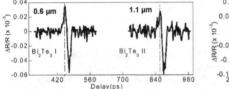 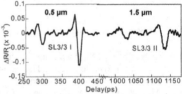

**Figure 7.** Coherent acoustic phonon signal for Bi$_2$Te$_3$ (left) and a SL with 3nm Bi$_2$Te$_3$ / 3nm Sb$_2$Te$_3$ (right). Film thicknesses are indicated. For increased thickness, the phonon reflectivity signal amplitude decreases significantly faster in the SL [45].

The effective longitudinal sound velocities calculated with the measured arrival times of acoustic echoes are ~ 2450 – 2550 m/s, lower than that of the bulk (Bi$_2$Te$_3$: 2600 m/s, Sb$_2$Te$_3$: 2900 m/s). To conclude, both phonon reflectivity amplitude and sound velocity are reduced by the nanostructuring, resulting in a lower thermal conductivity.

**b) Transport properties of films grown by MBE**

High-quality SLs with the system Bi$_2$Te$_3$–Bi$_2$(Se,Te)$_3$ were grown with the MBE technique and a complete in-plane thermoelectrical dataset was reported [47]. Like for the MOCVD films the in-plane thermal conductivity was clearly reduced, but only a 25% increase for the in-plane ZT was observed for the SL structures compared to the optimized component materials.

**c) Transport properties of films grown by sputtering**

Compared to MOCVD grown films, a co-sputtered Bi$_2$Te$_3$ / Sb$_2$Te$_3$ SL film with a single layer thickness of 10 nm [52] displayed a conductivity being about one order of magnitude smaller and also a lower

Seebeck coefficient of 105 μV/K. It can be assumed that the rather modest electrical properties are responsible for the rather modest $ZT$ of only 0.005. Consequently tests of such SLs as a micro-cooler demonstrated a very poor performance. Ion bombardment of such films significantly improved the thermoelectric performance [51]. The thermal conductivity decreased with increasing ion bombardmer intensity. For the highest ion current, a reduction by 60% was reported. Simultaneously, extremely hig Seebeck coefficients up to -800 μV/K (all films were n-type!) were measured, which was attributed to quantum-well confinement effects (see [14]) due to the nanoclusters formed by the ion bombardment. Notably, the electrical conductivity did not suffer from the ion treatment. A ZT of 0.8 was reached.

Experiments on nanoalloyed $Sb_2Te_3/(Bi,Sb)_2Te_3$ films deposited by sputtering showed high pow factors of over 40 μW/cmK² [53]. Very low thermal conductivities were achieved, and for the fil annealed at the lowest temperature of 150 °C, the thermal conductivity was reduced to values as low 0.45 W/mK due to the 2D-nanostructuring. However, it was not possible to combine this low therm conductivity with sufficiently good electric properties, i.e. the power factor after 150 °C annealing w only ~15 μW/cmK². For the best film annealed at 300 °C (PF~ 40 μW/cmK²), a ZT between 0.6 and 1 was estimated.

### d) Transport properties of films grown by electrochemical deposition

As mentioned above electrochemically deposited $Bi_2Te_3/Sb_2Te_3$ SLs were only partially characterized with respect to thermoelectric properties. Hence the potential of this deposition method for the fabrication of SLs with good thermoelectric performance cannot be judged at the moment.

### e) Transport properties of films grown by Pulsed Laser Deposition

Extremely low values for the thermal conductivity of ~ 0.11 W/mK (layer thickness of 6 nm) were achieved on symmetric $Bi_2Te_3/Sb_2Te_3$ SLs [57], which is even below the minimum thermal conductivities predicted for $Bi_2Te_3$ by the model of Slack and Cahill [17]. The authors did not measure electrical properties and propose that if the measured values are combined with bulk electrical properties which is a rather optimistic assumption, a $ZT$ of 1.5 may be estimated.

## CONCLUSIONS

An overview over the theoretical and experimental advances of $Bi_2Te_3/Sb_2Te_3$ and related superlattice systems was given in this work. Currently, the outstanding properties of superlattice structures reporter in 2001 still have not been reproduced until now, despite significant efforts undertaken by many group Other approaches such as MBE, sputtering, PLD and electrochemical deposition were applied, but complete sets of transport and structural properties are rarely presented. Values for the figures of merit comparable to those published in [17] and comparable structural properties (low dimension and therm: stability) have not been achieved until now.

## ACKNOWLEDGEMENT

The authors thank the Deutsche Forschungsgesellschaft (DFG) for funding this work via SPP 1386"Nanothermoelectrics."

## REFERENCES

[1] W. Haken, *Annalen der Physik* **1910**, *337*, 291
[2] E. Altenkirch, *Phys. Z.* **1911**, *12*, 920
[3] Personal communication with Prof. U. Birkholz. See also ref. [5].
[4] J. Goldsmid, *Proc. Phys. Soc. B* **1958**, *71*, 633
[5] U. Birkholz, *Thermoelektrische Stromerzeuger*, chapter in K. J. Euler, *Energiedirektumwandlung* (p. 62ff), Thiemig-Taschenbücher, Gießen **1967**
[6] E. Justi,G. Neumann, G. Schneider, *Zeitschrift für Physik* **1959**, *156*, 217
[7] K.C. Handel, *Anfänge der Halbleiterforschung und –entwicklung, Dargestellt an den Biographien von vier deutschen Halbleiterpionieren*, Ph.D. thesis at the Rheinisch-Westfälische Technische Hochschule Aachen, **1999**, p.152
[8] A.F. Ioffe, *Doklady Akademie nauk SSSR* **1956**, *106*, 981
[9] D.M. Rowe, *Modern thermoelectrics*, Holt, Rhinehart and Winston Ltd. (London), **1983**
[10] A.F. Joffe, L.S. Stil'bans, *Rep. Prog. Phys.* **1959**, *22*, 167.
[11] C.B. Vining, *Proceedings ICT 1992: 11ʰ International Conference on Thermoelectrics, Arlington, USA* **1992**, 223
[12] G.D. Mahan, *J. Appl. Phys.* **1989**, *65*, 1578
[13] G. Min, D.M. Rowe, *Appl. Phys. Lett* **2000**, *77*, 860
[14] L.D. Hicks, M.S. Dresselhaus, *Phys.Rev. B* **1993**, *47*, 12727
[15] B. Poudel, Q. Hao, Y. Ma, Y. Lan, A. Minnich, B. Yu, C. Yan, D. Wang, A. Muto, D. Vashaee, X. Chen, J. Liu, M.S. Dresselhaus, G. Chenm Z. Ren, *Science Express* **2008**, *320*, 634
[16] H. Böttner, D.G. Ebling, A. Jacquot, J. König, L. Kirste, J. Schmidt, *Phys. Status Solidi RRL* **2007**, *1*, 235
[17] R. Venkatasubramanian, E. Siivola, T. Colpitts, B. O'Quinn., *Nature* **2001**, *413*, 597
[18] L.D. Hicks, T.C. Harman, X. Sun, M.S. Dresselhaus, *Phys. Rev. B* **1996**, *53*, R10493
[19] P.Y. Yu, M. Cardona, *Fundamentals of Semiconductors – Physics and Materials Properties*, Springer-Verlag, Berlin Heidelberg, **1996**
[20] T. Koga, X. Sun, S.B. Cronin, M.S. Dresselhaus, *Appl. Phys. Lett.* **1998**, *73*, 2950
[21] A. Shakouri, Proceedings ICT 2005: 24ʰ International Conference on Thermoelectrics, Clemson, USA **2005**, 495
[22] K. Nielsch, J. Bachmann, J. Kimling, H. Böttner, *Adv. Energy Mater.* **2011**, *1*, 713
[23] J.M.O. Zide, D. Vashaee, Z.X. Bian, G. Zeng, J.E. Bowers, A. Shakouri, A.C. Gossard, *Phys. Rev. B* **2006**, *74*, 205335
[24] J.P. Heremans, C.M. Thrush, D. Morelli, *J. Appl. Phys.* **2005**, *98*, 063703
[25] N.F. Hinsche, B.Y. Yavorsky, I. Mertig, P. Zahn, *Phys.Rev. B* **2011**, *84*, 165214
[26] B.Y. Yavorsky, N.F. Hinsche, I.Mertig, P. Zahn, *Phys. Rev. B* **2011**, *84*, 165208
[27] N.F. Hinsche, B.Yu. Yavorsky, M. Gradhand, M. Czerner, M. Winkler, J. König, H. Böttner, I. Mertig, P. Zahn, *Phys.Rev.B* **2012**, *86*, 085323
[28] S. Tamura, D.C. Hurley, J.P. Wolfe, *Phys. Rev. B* **1988**, *38*, 1427
[29] V. Narayanamurti, H.L. Störmer, M.A. Chin, A.C. Gossard, W. Wiegmann, *Phys. Rev. Lett.* **1979**, *43*, 2012
[30] G. Chen, *Phys. Rev. B* **1998**, *57*, 14958
[31] P. Hyldgaard, G.D. Mahan, *Phys. Rev. B* **1997**, *56*, 10754
[32] S.Y. Ren, J.D. Dow, *Phys. Rev. B* **1982**, *25*, 3750
[33] R. Venkatasubramanian, *Phys. Rev. B* **2000**, *61*, 3091
[34] A. Pattamatta, C.K. Madnia, *Int. J. Heat. Mass. Tran.* **2009**, *52*, 860
[35] F. Takahashi, Y. Hamada, T. Mori and I. Hatta, *Jpn. J. Appl. Phys.* **2004**, *43*, 8325
[36] R. Venkatasubramanian, T. Colpitts, E. Watko, J. Hutchby, *Proceedings ICT 1996: 15ʰ International Conference on Thermoelectrics, Pasadena, USA* **1996**, 454
[37] R. Venkatasubramanian, T. Colpitts, E. Watko, M. Lamvik, N. El-Masry, *J. Cryst. Growth* **1997**, *170*, 817

[38] R. Venkatasubramanian, T. Colpitts, B. O'Quinn, M. Lamvik, N. El-Masry, *Appl. Phys. Lett.* **1999**, *75*, 1104

[39] M.N. Touzelbaev, P. Zhou, R. Venkatasubramanian, K. E. Goodson, *J. Appl. Phys.* **2001**, *90*, 763

[40] H. Cui, I. Bhat, B. O'Quinn, R. Venkatasubramanian, *J. Electron. Mater.* **2001**, *30*, 1376

[41] K.D. Coonley, B.C. O'Quinn, J.C. Caylor, R. Venkatasubramanian, *Mat. Res. Soc. Symp. Proc.* **2004**, *793*, S2.5.1

[42] R. Venkatasubramanian, E. Siivola, B. O'Quinn in *CRC Handbook of Thermoelectrics (Ed. D.M. Rowe)*, CRC Press, Boca Raton, **2006**, 49-1

[43] G.E. Bulman, E. Siivola, B. Shen, R. Venkatasubramanian, *Appl. Phys. Lett* **2006**, *89*, 122117

[44] Y. Wang, X. Xu, R. Venkatasubramanian, *Appl. Phys. Lett.* **2008**, *93*, 113114

[45] Y. Wang, C. Liebig, X. Xu, R. Venkatasubramanian, *Appl. Phys. Lett.* **2010**, *97*, 083103

[46] Y. Iwata, H. Kobayashi, S. Kikuchi, E. Hatta, K. Mukasa, *J. Cryst. Growth* **1999**, *203*, 125

[47] J. Nurnus, H. Böttner, A. Lambrecht in *CRC Handbook of Thermoelectrics (Ed. D.M. Rowe)*, CRC Press, Boca Raton, **2006**, 48-1

[48] J.D. König, M. Winkler, S. Buller, W. Bensch, U. Schürmann, L. Kienle, H. Böttner, *J. Electron. Mater.* **2011**, *40*, 1266

[49] M. Winkler, J. D. König, S. Buller, U. Schürmann, L. Kienle, W. Bensch, H. Böttner, *MRS proceedings (spring meeting 2011)* **2011**, 1329, mrss11-1329-i10-15

[50] C. Mortensen, R. Rostek, B. Schmid, D.C. Johnson, *Proceedings ICT 2005: 24th International Conference on Thermoelectrics*, Clemson, USA **2005**, 261

[51] B. Zheng, Z. Xiao, B. Chhay , R.L. Zimmerman, Matthew E. Edwards , D. ILA, *Surface & Coatings Technology* **2009**, *203*, 2682

[52] Z. Xiao, K. Hedgemen, M. Harris, E. DiMasi, *J. Vac. Sci. Technol. A* **2010**, *28*, 679

[53] M. Winkler, X. Liu, J. D. König, L. Kirste, H. Böttner, W. Bensch, L. Kienle, *J. Electron. Mater.* **2012**, *41*, 1322

[54] M.A. Ryan, J.A. Herman, J.-P. Fleurial, *206th Meeting 2004 of The Electrochemical Society, Inc* **2004**, Abs. 238

[55] W. Zhu , J.-Y. Yang, D.-X. Zhou, C.-J. Xiao, X.-K. Duan, *Langmuir* **2008**, *24*, 5919

[56] D. Banga, J. L. Lensch-Falk, D.L. Medlin, V. Stavila, N.Y.C. Yang, D.B. Robinson, P.A. Sharma, *Cryst. Growth Des.* **2012**, *12*, 1347

[57] I. Yamasaki, R. Yamanaka, M. Mikami, H. Sonobe, Y. Mori, T. Sasaki, *Proceedings ICT 1998: 17th International Conference on Thermoelectrics*, Nagoya, Japan **1998**, 210

[58] G. Slack in *CRC Handbook of Thermoelectrics (Ed. D.M. Rowe)*, CRC Press, Boca Raton, **1995**, chap. 34, 407

[59] D. Teweldebrhan, V. Goyal, A. A. Balandin, *Nano Lett.* **2010**, *10*, 1209

[60] V. Goyal, D. Teweldebrhan, A.A. Balandin, *Appl. Phys. Lett.* **2010**, *97*, 133117

[61] S. Nakajima, *J. Phys. Chem. Solids* **1963**, *24*, 479.

[62] T. L. Anderson, H. B. Krause, *Acta Crystallogr., Sect. B: Struct. Crystallogr. Cryst. Chem.* **1974**, *30*, 1307

[63] H. Scherrer, S. Scherrer in *CRC Handbook of Thermoelectrics (Ed. D.M. Rowe)*, CRC Press, Boca Raton, **1995**, chap. 19, 211

[64] G. Wang, L. Endicott, C. Uher, *Sci. Adv. Mater.* **2011**, *3*, 539

[65] J. Nurnus, *Thermoelektrische Effekte in Übergittern und Multi-Quantentrog-Strukturen*, PhD thesis at University of Freiburg, Germany, 2001

[66] J. Nurnus et al., *Proceedings ICT 2000: 19th International Conference on Thermoelectrics*, Cardiff, UK **2000**, 236

[67] Personal information from Guoyo Wang on epitaxial Sb$_2$Te$_3$ films, talk on International Conference on Thermoelectrics, Traverse City, USA, 2011

[68] N. Peranio, O. Eibl, J. Nurnus, *J. Appl. Phys.* **2006**, *100*, 114306

[69] Z. Aabdin, N. Peranio, O.Eibl, *Adv. Mater.* **2012**, DOI: 10.1002/adma.201201079

[70] N. Peranio, O. Eibl, *J. Appl. Phys.* **2008**, *103*, 024314

[71] N. Peranio, O. Eibl, *Phys. Status Solidi A* **2009**, *1*, 42
[72] U. Schürmann, M. Winkler, J.D. König, X. Liu, V. Duppel, W. Bensch, H. Böttner, L. Kienle, *Adv. Eng. Mater.* **2012**, *14*, 139
[73] M. Winkler, Jan D. König, S. Buller, U. Schürmann, L. Kienle, W. Bensch, H. Böttner, *Proceedings of the 8th European Conference on Thermoelectrics*, Como, Italy **2010**, 19
[74] C.H. Champness, P.T. Chiang, P. Parker, *Can. J. Phys.* **1965**, *43*, 653
[75] J.R. Sootsman, D.Y. Chung, M.G. Kanatzidis, *Angew. Chem. Int. Ed.* **2009**, *48*, 8616

Mater. Res. Soc. Symp. Proc. Vol. 1490 © 2012 Materials Research Society
DOI: 10.1557/opl.2012.1573

Impact of Annealing on the Thermoelectric Properties of Ge$_2$Sb$_2$Te$_5$ Films

Jaeho Lee, Takashi Kodama, Yoonjin Won, Mehdi Asheghi, Kenneth E. Goodson
Department of Mechanical Engineering, Stanford University, Stanford, California 94305, U.S.A.

ABSTRACT

Thermoelectric phenomena strongly influence the behavior of chalcogenide materials in nanoelectronic devices including phase-change memory cells. This paper presents the annealing temperature and phase dependent thermoelectric properties of Ge$_2$Sb$_2$Te$_5$ films including the thermoelectric power factor and the figure of merit. The Ge$_2$Sb$_2$Te$_5$ films annealed at different temperatures contain varying fractions of the amorphous and crystalline phases which strongly influence the thermoelectric properties. The thermoelectric power factor increases fom 3.2 $\mu$W/mK$^2$ to 65 $\mu$W/mK$^2$ as the crystal phase changes from face-centered cubic to hexagonal close-packed. The data are consistent with modeling based on effective medium theory and suggest that careful consideration of phase purity is needed to improve the figures of merit for phase change memories and potentially for thermoelectric energy conversion applications.

INTRODUCTION

Thermoelectric transport can have a large impact on the performance of semiconductor devices and related nanostructures [1-4]. The impact is possibly most pronounced in phase-change memories [5], which experience both large current densities and temperature excursions exceeding 600 °C. Recent measurements provided evidence of thermoelectric transport in phase-change cells by capturing a modification in the amorphous region [1] and the programming condition [2] with the bias polarity. However, there are relatively few data for the thermoelectric properties of phase-change materials at the film thicknesses relevant for contemporary devices.

Past measurements of various chalcogenide materials revealed that holes are responsible for their positive Seebeck coefficient and the Seebeck coefficient strongly depends on phase and processing conditions [6-8]. However, the existing data for the Seebeck coefficient did not capture the impact of annealing and its impact on phase purity while the Ge$_2$Sb$_2$Te$_5$ films can often contain a mixture of amorphous and crystalline phases, and the phase purity can strongly influence transport phenomena [9]. Our recent work reported the Seebeck coefficient of 25-nm and 125-nm-thick Ge$_2$Sb$_2$Te$_5$ films between room temperature and 300 °C using an experimental structure in which a buried oxide layer induces lateral temperature fields [10]. Here we describe the impact of annealing on the thermoelectric properties of Ge$_2$Sb$_2$Te$_5$ including the figure of merit for thermoelectric applications. The phase-dependent Seebeck coefficient is explained by different models, including the effective medium theory for treating phase impurities.

EXPERIMENTAL DETAILS

Samples are prepared on an SOI substrate with a 1.5-$\mu$m-thick silicon layer and a 1-$\mu$m-thick buried oxide layer. A 70 nm-thick silicon dioxide film is deposited on the SOI substrate by plasma-enhanced chemical vapor deposition for electrical passivation. Electrodes are then patterned by e-beam lithography using a 300-nm-thick PMMA resist. A 150-nm-thick platinum film with a 2-nm-thick titanium adhesion layer is deposited by sputtering and lifted off in an acetone bath. Another layer of e-beam lithography produced patterns for Ge$_2$Sb$_2$Te$_5$ structures with 0.5 μm alignment accuracy. The Ge$_2$Sb$_2$Te$_5$ films are deposited by DC magnetron sputtering in argon with a pressure of 2 mT at room temperature and lifted off in an acetone bath. While the as-deposited Ge$_2$Sb$_2$Te$_5$ is amorphous, crystalline structures in face-centered cubic (fcc) and hexagonal close-packed (hcp) phases are obtained by annealing the samples at elevated temperatures.

The experimental strategies utilizing the thermal healing length on the SOI substrate (Fig.1) are described elsewhere [10]. Measuring the hot junction temperature ($\Delta T$) and the open-circuit voltage ($\Delta V$) across the Ge$_2$Sb$_2$Te$_5$ structure provides the Seebeck coefficient ($S = \Delta V/\Delta T$). The Seebeck coefficient and the electrical resistivity are successively measured at each temperature controlled by a hot chuck. The electrical resistivity and the Seebeck coefficient of Ge$_2$Sb$_2$Te$_5$ films are measured by following the same pre-annealing and ramping details. Each pre-annealing condition produces a unique fraction of crystals and impurities, verified by the X-ray diffraction (XRD) data and the Johnson–Mehl–Avrami–Kolmogrov (JMAK) theory. The phase and temperature dependent Seebeck coefficient data correspond well with the trend in the electrical resistivity data. The effects of pre-annealing indicate that the both properties are strongly dependent on the phase quality.

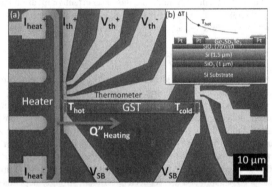

FIG. 1. Top-view SEM image (a) and a cross-sectional schematic (b) of the experimental structure used here for measuring the Seebeck coefficient of Ge$_2$Sb$_2$Te$_5$ films. The platinum heater, far left, provides uniform heating (Q") and the temperature gradient necessary for the Seebeck voltage measurements. The optimized thermal healing length of the SOI substrate minimizes temperature rise in the cold junction.

**DISCUSSION**

Phase impurities in chalcogenide materials can complicate the prediction of the Seebeck coefficient and require dedicated modeling using, for example, the effective medium theory, as developed later in this section. The classical models, however, with accurate predictions of phase using the XRD measurements can properly capture the temperature dependence of Seebeck coefficient. Our data show that the Seebeck coefficients of amorphous and fcc Ge$_2$Sb$_2$Te$_5$ films decrease with increasing temperature [10]. In the case of a homogeneous non-degenerate semiconductor, the Seebeck coefficient can be approximated using [7, 11],

$$S = \frac{k_B}{e}\left(\frac{E_S}{k_B T}+A\right) = \frac{k_B}{e}\left(\ln\left(\frac{N_v}{p}\right)+A\right)$$ (1)

where $k_B$ is the Boltzmann constant, $A$ is the transport constant, $e$ is the carrier charge, and $E_s$ is the activation energy, which is related to the carrier density $p = N_v exp(-E_s/k_B T)$. Increased annealing temperature induces crystallization that activates more carriers, and the increased carrier density results in the decreased Seebeck coefficient (Fig. 2).

The Seebeck coefficient and the electrical resistivity of hcp $Ge_2Sb_2Te_5$ films both increase with increasing temperature, indicating that the material is degenerate semiconductor. The Mott formula approximates the Seebeck coefficient in the frame work of nearly free-electrons [11, 12],

$$S = \frac{\pi^2 k_B^2 T}{3eE_F}(1+U) = \frac{\pi k_B^2 T}{6eh^2}\left(\frac{8\pi}{3p}\right)^{2/3}(1+U)$$ (2)

where $U$ is a constant that depends on the scattering mechanism, $E_F$ is the difference between the Fermi energy and the band edge. This approximation is valid for spherical Fermi surface that is independent of temperature, as in the free-electron model. Again with larger annealing temperature, the Seebeck coefficient decreases due to increased carrier density (Fig. 2). Increased annealing temperature induces crystallization that activates more carriers, and the increased carrier density results in the decreased Seebeck coefficient (Fig. 2). While the Seebeck coefficient models (Eq. 1-2) are based on a single-phase assumption, the $Ge_2Sb_2Te_5$ films pre-annealed at 130 °C and at 300 °C may contain significant phase impurities. A reasonable way to approximate the Seebeck coefficient of a composite material with different phases is to use the effective medium theory (EMT) [10, 11],

$$\sum_i v_i \frac{k_i/S_i - k_{eff}/S_{eff}}{k_i/S_i + 2k_{eff}/S_{eff}} = 0$$ (3)

The parameters $v_i$, $k_i$, and $S_i$ are the volume fraction, the thermal conductivity, and the Seebeck coefficient of a phase $i$, and $k_{eff}$ and $S_{eff}$ are the effective properties. The EMT captures the effective Seebeck coefficient and the effective thermal conductivity of $Ge_2Sb_2Te_5$ film as a function of crystal fraction, which corresponds to the phase purity content. The EMT matches the Seebeck coefficient data for the $Ge_2Sb_2Te_5$ film pre-annealed 130 °C to the crystal fraction of ~0.25, which is consistent with the XRD data. The EMT model also predicts that the hcp phase is dominant in the $Ge_2Sb_2Te_5$ composites that are pre-annealed at 300 °C.

Another thermoelectric property relevant for electronic devices is the Thomson coefficient ($\mu_T = T \partial S/\partial T$), which governs heat absorption or release in a medium carrying an electrical current. The Thomson coefficient of 25-nm-thick $Ge_2Sb_2Te_5$ ranges from -400 to -500 µV/K in the amorphous phase, -50 to -150 µV/K in the fcc phase, and from +15 to +60 µV/K is in the hcp phase. The Thomson coefficient becomes positive as the material becomes a degenerate semiconductor with pre-annealing temperatures greater than 300 °C.

Thermoelectric materials and their properties have drawn attention mostly for its ability to convert heat into electricity. The dimensionless figure of merit ($ZT = S^2 T/k\rho$) is expected to be large for phase change materials because they offer excellent electrical conduction with poor thermal conduction, which is closely related to the phase change memory device figure of merit. Figure 3 shows the thermoelectric power factor and the thermoelectric figure of merit for 25-nm-thick $Ge_2Sb_2Te_5$ films that are annealed at varying temperatures. The $ZT$ of $Ge_2Sb_2Te_5$ films is highest in the fcc phase due to the large thermoelectric power and the relatively low thermal conductivity [9]. The $ZT$ of fcc $Ge_2Sb_2Te_5$ films will increase with increasing temperature because the Seebeck coefficient decreases with temperature, as predicted by Equation 1. For the same reason, the $ZT$ of hcp $Ge_2Sb_2Te_5$ films will increase with increasing temperature as predicted by Equation 2. Because the electron contribution to the thermal conductivity is coupled with the electrical conductivity by the Wiedemann-Franz law, the phonon contribution of the thermal conductivity needs to be tuned for improving the thermoelectric performance.

Scattering of carriers on material surface and grain boundaries can strongly influence the transport [13, 14]. However, the carrier mean free path ($\lambda$) in crystalline $Ge_2Sb_2Te_5$ is much smaller than the film thicknesses or the grain size. A simple estimate for the mean free path using $\lambda = v_{th}\tau$ predicts that the carrier mean free path in $Ge_2Sb_2Te_5$ is only a few nanometers. The grain size estimated by XRD peak broadening is about 20 nm for both the 25-nm and the 125-nm-thick films. Neither the grain boundary scattering nor the surface scattering contributes to the thermoelectric properties of $Ge_2Sb_2Te_5$ films. While other thermoelectric materials improve their $ZT$ through geometric scaling, the thermoelectric properties of $Ge_2Sb_2Te_5$ films show very weak dependence on the film thickness down to 25 nm due to limited scattering on surface of material boundaries. Our recent work on the Seebeck coefficient [10] argues that thickness dependent crystallization effect and the resultant carrier density are responsible for the reduction in the Seebeck coefficient of the 25-nm-thick films, compared to that of the 125-nm-thick films. This indicates that the thermoelectric figure of merit can be further improved by carefully controlling the annealing temperature and phase purity in $Ge_2Sb_2Te_5$ films.

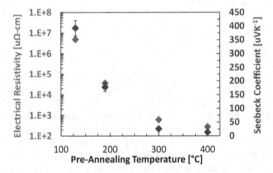

FIG. 2. Impact of annealing on the electrical resistivity and the Seebeck coefficient of 25-nm-thick $Ge_2Sb_2Te_5$ films. Previously reported data[10,11] for the pure amorphous phase and the fully crystalline phase provide constraints for matching the data for the $Ge_2Sb_2Te_5$ film pre-annealed 130 °C to the crystal fraction ~0.25. This result is consistent with the XRD analysis in Figure 3.

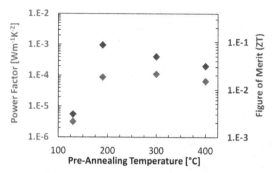

FIG. 3. Impact of annealing on the thermoelectric power factor and the thermoelectric figure of merit of 25-nm-thick Ge$_2$Sb$_2$Te$_5$ films. The thermoelectric power factor (S$^2$/ρ) calculated from the data in Figure 6 is larger in the crystalline phases despite the reduction in the Seebeck coefficient. The thermoelectric figure of merit (S$^2$T/ρ/k) estimated using the known thermal conductivity data [11] shows the peak value in the fcc phase due to low thermal conductivity.

## CONCLUSIONS

The thermoelectric properties of Ge$_2$Sb$_2$Te$_5$ films show strong dependence on temperature history as governed by the phase purity. The high thermoelectric figure of merit reported in the fcc phase show that Ge$_2$Sb$_2$Te$_5$ films can be a good candidate for thermoelectric energy conversion applications with careful control of annealing temperature and phase purity. For electronic device applications, the thermoelectric properties identified in this study improve the quality of simulations that provide a detailed view of temperature distributions. Precise knowledge about thermoelectric transports in Ge$_2$Sb$_2$Te$_5$ films can thus allow the development of viable design strategies for novel phase-change memories.

## ACKNOWLEDGMENTS

The Stanford authors appreciate support from Intel Corporation, the Semiconductor Research Corporation through contract 2009-VJ-1996, and the National Science Foundation through grant CBET-0853350. The authors acknowledge the Stanford Nanofabrication Facility (SNF) for supports in fabrication.

## REFERENCES

[1] D. T. Castro, L. Goux, G. A. M. Hurkx, K. Attenborough, R. Delhougne, J. Lisoni, F. J. Jedema, M. A. A. t Zandt, R. A. M. Wolters, and D. J. Gravesteijn, "Evidence of the thermo-electric Thomson effect and influence on the program conditions and cell optimization in

phase-change memory cells," in *IEDM Tech. Dig.*, pp. 315–318, 2007.

[2] D.-S. Suh, C. Kim, K. H. P. Kim, Y.-S. Kang, T.-Y. Lee, Y. Khang, T. S. Park, Y.-G. Yoon, J. Im, J. Ihm, "Thermoelectric heating of $Ge_2Sb_2Te_5$ in phase change memory devices," *Appl. Phys. Lett.*, vol.96, no.12, Mar 2010.

[3] K. L. Grosse, M.-H. Bae, F. Lian, E. Pop, W. P. King, "Nanoscale Joule Heating, Peltier Cooling and Current Crowding at Graphene-Metal Contacts," Nature Nano 2011, 6, 287.

[4] D.-K. Kim, and Yoondong Park, "Polarity-Dependent Morphological Changes of Ti/TiN/W Via Under High Current Density," *Electron Device Letters, IEEE* , vol.31, no.2, pp.120-122, Feb. 2010.

[5] J. Lee, M. Asheghi, and K. E. Goodson, "Impact of thermoelectric effects on phase-change memory performance metrics and scaling," *Nanotechnology*, 23, 205201, 2012.

[6] F. Yan, T.J. Zhu, X.B. Zhao, and S.R. Dong, "Microstructures and thermoelectric properties of GeSbTe based layered compounds," *Appl. Phys. A* 88, 425, 2007.

[7] S. A. Baily, D. Emin, and H. Li, "Hall mobility of amorphous Ge2Sb2Te5," *Solid State Commun.*, 139 161, 2006.

[8] T. Kato and K. Tanaka, "Electronic properties of amorphous and crystalline Ge2Sb2Te5 films," *Jpn. J. Appl. Phys.*, vol. 44, no. 10, pp. 7340–7344, Oct. 2005.

[9] J. Lee, Z. Li, J. P. Reifenberg, M. Asheghi, and K. E. Goodson, "Thermal Conductivity Anisotropy and Grain Structure in $Ge_2Sb_2Te_5$ Films," *Journal of Applied Physics*, vol. 109, 084902, Apr. 2011.

[10] J. Lee, T. Kodama, Y. Won, M. Asheghi, and K. E. Goodson, "Phase Purity and the Thermoelectric Properties of $Ge_2Sb_2Te_5$ Films down to 25 nm Thickness," Journal of Applied Physics, 112, 014902, Jul. 2012

[11] J. Sonntag, "Thermoelectric power in alloys with phase separation (composites)," *Phys. Rev. B*, 73, 045126, 2006.

[12] J. M. Ziman, "Electrons and Phonons," *Clarendon, Oxford*, p. 397, 1962

[13] C. R. Tellier and A. J. Tosser, "THERMOELECTRIC POWER OF METALLIC FILMS IN THE MAYADAS-SHATZKES MODEL," *Thin Solid Films* 41, 161-166 (1977).

[14] C. R. Pichard, C. R. Tellier, and A. J. Tosser, "Thermoelectric power of thin polycrystalline metal films in an effective mean free path model," *J. Phys. F: Metal Phys.* 10, 2009-14 (1980).

Mater. Res. Soc. Symp. Proc. Vol. 1490 © 2012 Materials Research Society
DOI: 10.1557/opl.2012.1558

# Structural and Electrical Properties of Mg–Si Thin Films Fabricated by Radio-Frequency Magnetron Sputtering Deposition

Jun-ichi Tani and Hiroyasu Kido
Osaka Municipal Technical Research Institute, 1-6-50 Morinomiya Joto-ku Osaka 536-8553, Japan

## ABSTRACT

Mg–Si thin films were fabricated on glass, Si(100), Si(111), and polycrystalline $Al_2O_3$ substrates by radio-frequency (RF) magnetron sputtering deposition using an elemental composite target composed of Si chips on a Mg disk. The effect of deposition conditions such as the composition ratio of Mg/Si in the target area, substrate temperature, and the type of substrate on the thin film deposition was investigated. By controlling the deposition conditions, pure-phase $Mg_2Si$ polycrystalline films were successfully fabricated at room temperature. The crystalline orientation of the films was strongly influenced by the Mg/Si elemental composition ratio in the targets as well as by the surface roughness and porosity of the substrate. The electron concentration and mobility of nondoped $Mg_2Si$ films were $2.2 \times 10^{16}$ cm$^{-3}$ and 2.0 cm$^2$/Vs, respectively. The electron concentration of $Mg_2Si$ films was drastically increased by impurity doping with Al and Bi.

## INTRODUCTION

Magnesium silicide ($Mg_2Si$) is an environmentally friendly semiconductor with an indirect band gap of 0.6–0.8 eV [1-3]. This material has attracted much attention over the past few years in view of its potential application in thermoelectric generators, solar photovoltaic (PV) cells, thermophotovoltaic (TPV) cells, infrared (IR) sensors, and structural applications. The raw materials are inexpensive because Si is the second most abundant element in the Earth's crust and Mg is the eighth most abundant. Moreover, $Mg_2Si$ has a number of advantageous features: it is nontoxic, lightweight, has high electron mobility and high optical absorption coefficients, and demonstrates good thermoelectric performance [4, 5].

Many researchers have focused on bulk materials rather than thin films because it has been pointed out that the fabrication of $Mg_2Si$ thin films is difficult because of the very high vapor pressure of Mg at temperatures as low as 473 K and the low condensation coefficient of Mg on a Si surface [6]. In general, magnetron sputtering has been widely used in industrial production processes for large-area high-quality thin-film deposition. However, only few reports have been published on using magnetron sputtering technique for the preparation $Mg_2Si$ thin films. Hasapis et al. [7] reported the preparation of $Mg_2Si$ thin films on Si substrates by dual cathode magnetron sputtering (DSMS). Xiao et al. [8] reported the synthesis of $Mg_2Si$ films on silicon (111) substrates by RF magnetron sputtering deposition and subsequent annealing. Yamaguchi et al. [9] reported the deposition of Mg–Si films on AZ31 magnesium alloys by a high-frequency sputtering method using a $Mg_2Si$ target. Kato et al. [10] reported the fabrication of polycrystalline $n$-type $Mg_2Si$ semiconducting layers with a uniform grain size of approximately 50 nm by applying RF magnetron sputtering using a $Mg_2Si$ target at a high sputtering gas pressure ($Ps$) of 9 Pa and a substrate temperature ($Ts$) of 473 K. However, the effect of sputtering target composition on the structure and electrical properties of the Mg–Si films fabricated by RF magnetron sputtering deposition have not been investigated in detail.

In the present study, we fabricated Mg–Si thin films on glass, Si(100), Si(111), and polycrystalline porous $Al_2O_3$ substrates by RF magnetron sputtering deposition using an elemental composite target composed of Si chips on a Mg disk. The effect of deposition conditions such as Mg/Si elemental area ratio of the target, sputtering power, type of substrate, $Ts$, and $Ps$ were investigated. The electrical properties of the films were investigated by Hall effect measurements. Moreover, impurity-doped Mg–Si thin films were successfully fabricated using a three-element composite target composed of Mg, Si, and impurities (Al or Bi).

**EXPERIMENT**

Mg–Si films were deposited by RF magnetron sputtering (sputtering power: 300 W; model: HSR-551S, Shimadzu Corporation, Kyoto, Japan) for 30 min under an Ar atmosphere ($Ps$ = 0.667 Pa) on glass (S-1111, crown glass, Matsunami Co. Ltd., Japan), Si(100) wafer (n-type, electrical resistivity ($\rho$): 8.3 $\Omega$cm), Si(111) wafer (n-type, $\rho$: 2.0 $\Omega$cm), and polycrystalline $Al_2O_3$ substrates (purity: 96%, density: 3.7 g/cm$^3$, Mitani Micronics Kyushu Co. Ltd., Japan). The substrate temperature ($Ts$) ranged from room temperature (RT) to 573 K. The sputtering target was composed of Si chips (purity: >99.999%; Size: 10 × 10 mm, Kojundo Chemical Laboratory Co., Ltd.; Saitama, Japan) on a Mg disk (purity: >99.9%; diameter: 101.6 mm, Kojundo Chemical Laboratory Co., Ltd.; Saitama, Japan). The elemental composition of the Mg–Si films was controlled by changing the number of Si chips on the Mg disk. In order to fabricate the impurity-doped Mg–Si films, Al chips (purity: >99.99%; Size: 5 × 5 mm, Kojundo Chemical Laboratory Co., Ltd.; Saitama, Japan) or Bi chips (purity: >99.99%; Size: 5 × 5 mm, Kojundo Chemical Laboratory Co., Ltd.; Saitama, Japan), as well as Si chips on Mg disk, were used as a sputtering target.

The surface morphologies and elemental analyses of the films were characterized using scanning electron microscopy and energy-dispersive X-ray spectrometry (SEM-EDX; models JSM-6460LA and JSM-6610LA, JEOL, Tokyo, Japan) at 20 kV and a working distance of 10 to 20 mm. The samples were coated with a thin layer of conductive carbon, which allowed for the detection of carbon by EDX. Phase analysis was carried out by X-ray powder diffraction (XRD; model RINT 2500, Rigaku, Tokyo, Japan), utilizing Cu Kα radiation at 40 kV and 100 mA. Phase identification was accomplished by comparing the experimental XRD patterns with the standards compiled by the International Center for Diffraction Data (ICDD, Newtown Square, PA). The Hall coefficients ($R_H$) and $\rho$ at RT were measured by the van der Pauw method using a Toyo Corp. ResiTest 8320. The carrier concentration of the samples was determined by the factor $1/e|R_H|$.

**RESULTS AND DISCUSSION**

The Mg/Si atomic composition ratio of Mg–Si films was evaluated by EDX using a stoichiometric sintered $Mg_2Si$ bulk sample as a standard reference material. When the Mg/Si atomic composition ratio of the Mg–Si films was 2:1, which is the stoichiometric composition of the $Mg_2Si$ phase, the target area ratio of Mg/(Mg + Si) was 75.8%.

Figure 1 shows XRD patterns of the Mg–Si films deposited on (a) glass substrates and (b) polycrystalline $Al_2O_3$ substrates at $Ts$ = RT. When the sputtering target area ratio of Mg/(Mg +

Si) was 77.1–93.6%, XRD patterns of the Mg–Si films revealed the presence of Mg$_2$Si (PDF #35-0773, cubic), as well as Mg (PDF #35-0821, hexagonal). When the sputtering target area ratio was 68.2–75.8%, only Mg$_2$Si was present in the Mg–Si films, suggesting that pure-phase Mg$_2$Si polycrystalline films were successfully fabricated at $Ts$ = RT. Kato et al. [10] fabricated Mg–Si films by RF magnetron sputtering using a Mg$_2$Si target. They pointed out that the structure of the Mg$_2$Si layer deposited at $Ts$ = RT was mainly amorphous. The discrepancy between our results and those of Kato et al. can be explained by the different sputtering conditions. The sputtering conditions in this study (power =300 W and $Ps$ = 0.667 Pa) were totally different from their experimental conditions (power = 20 W and $Ps$ = 9.0 Pa). Yamaguchi et al. [9] pointed out that the structure in Mg–Si films was affected by $Ps$; an amorphous structure is formed at a $Ps$ over 4 Pa, whereas the Mg$_2$Si crystalline structure is formed at lower pressures. When the sputtering target area ratio was 49.1–61.8%, XRD patterns of the Mg–Si films revealed the presence of hexagonal Mg$_2$Si (PDF #34-0673), as well as cubic Mg$_2$Si (PDF #35-0773). When the sputtering target area ratio was 35.1%, no diffraction peaks appeared in the XRD patterns except the crystalline peaks of the Al$_2$O$_3$ substrate, results which reveal that Mg–Si films have an amorphous structure.

For the sputtering target area ratio of 77.1%, the strongest XRD peak was the (111) plane of the Mg$_2$Si phase on both the polycrystalline Al$_2$O$_3$ and the glass substrates. For the sputtering target area ratio of 75.8%, however, the strongest XRD peaks of the Mg$_2$Si phase for the glass and polycrystalline Al$_2$O$_3$ substrates were the (200) and (220) planes, respectively. The results suggest that the orientation of the Mg$_2$Si phase in the films depends on the elemental composition of the Mg–Si films. As discussed above, a small amount of Mg was detected in the XRD pattern of the Mg–Si film for the sputtering target area ratio of 77.1%. Therefore, a small

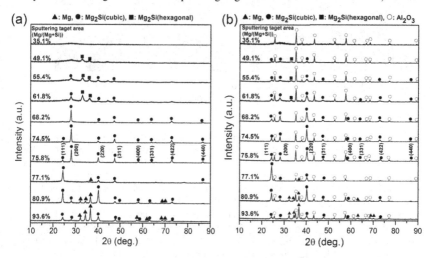

**Figure 1.** XRD patterns of the Mg–Si films deposited on (a) glass substrates and (b) polycrystalline Al$_2$O$_3$ substrates at $Ts$ = RT when the sputtering target area ratio of Mg/(Mg + Si) was 75.8%.

amount of Mg will affect the orientation of the Mg₂Si phase in the films. From the broadening of the most intense (200) peak in the Mg–Si films on glass from the sputtering target area ratio of 75.8%, the average crystalline size of Mg₂Si was estimated to be approximately 38 nm by the Debye–Sherrer equation and Warren's correction to account for instrumental broadening [11]. The Mg/Si target area dependence of the Mg₂Si phase orientation of Mg–Si films deposited on glass was the same as that on the Si(100) and Si(111) substrates. Because the relative density of the polycrystalline Al₂O₃ substrate was 92.5%, the orientation of the Mg₂Si phase in the films will be affected by the surface roughness and porosity of the substrate. The orientation of the Mg₂Si phase in the films was also influenced by $Ts$. When Mg–Si films were deposited at RT, 423 K, 473 K, and 573 K, the intensity of the (220) plane of the Mg₂Si phase gradually increased with increasing $Ts$. At $Ts$ > 473 K, many Mg₂Si crystals a few hundred nanometers in size were observed in the films.

Figure 2 shows SEM micrographs of the Mg–Si films deposited on glass and Al₂O₃ substrates at $Ts$ = RT for the sputtering target area ratio of 75.8%. The film thickness was 2.6 μm. The surface of the Mg–Si films deposited on glass was smooth. However, the surface of the Mg–Si films deposited on the Al₂O₃ substrate was relatively rough, as the result of many grains a few micrometers in size, which is the same as the grain size of the polycrystalline Al₂O₃ substrates.

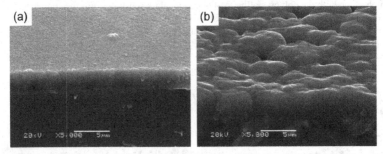

**Figure 2.** XRD patterns of the Mg–Si films deposited on (a) glass substrates and (b) polycrystalline Al₂O₃ substrates at $Ts$ = RT when the sputtering target area ratio of Mg/(Mg + Si) was 75.8%.

Figure 3 shows backscattered SEM micrographs of the surfaces of Mg–Si films deposited on Si(100), Si(111), and glass substrates at $Ts$ = RT for the sputtering target area ratio of 75.8%. We observed the generation of many defects, which were connected with each other, at the surface of the deposited Mg–Si films. The number of defects on the surfaces of the Mg–Si films deposited on Si(100) and Si(111) substrates was more than that of films deposited on glass substrates. The defects may be caused by the internal stress of the films resulting from the mismatch in the thermal expansion between the Mg–Si films and the substrates. The coefficient of thermal expansion (CTE) of polycrystalline Mg₂Si in the temperature range of 298–573 K is $16.5 \times 10^{-6}$ $K^{-1}$ [12]; this value is much higher than the CTE of Si ($2.6 \times 10^{-6}$ $K^{-1}$ at RT) [13] and glass ($10 \times 10^{-6}$ $K^{-1}$ at 373–653 K) [14]. When the substrate temperature was increased to 423–573 K, the generation of small cracks between the grain boundaries was observed in the films. Therefore, further optimization of the sputtering conditions is necessary in order to obtain defect-free Mg–Si thin films.

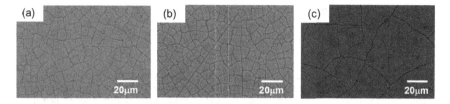

**Figure 3.** Backscattered SEM micrographs of the surfaces of Mg-Si films deposited on (a) Si(100), (b) Si(111), and (c) glass substrates at $Ts$ = RT when the sputtering target area ratio of Mg/(Mg + Si) was 75.8%.

Table 1 shows the transport properties of nondoped and Al- and Bi-doped $Mg_2Si$ films deposited on glass. The sign of $R_H$ for all $Mg_2Si$ films (Sample Nos. 1–7) is negative, indicating that the conductivity is mainly due to electrons. The electron concentration and mobility of nondoped Mg–Si films (1) are $2.2 \times 10^{16}$ $cm^{-3}$ and 2.0 $cm^2/Vs$, respectively. The mobility is much lower than the reported values for a polycrystalline bulk sample (204 $cm^2/Vs$ at $4.3 \times 10^{17}$ $cm^{-3}$) [15] and a thin film (24 $cm^2/Vs$ at $5.1 \times 10^{16}$ $cm^{-3}$) [10]. Because the estimated crystalline size (approximately 38 nm) of $Mg_2Si$ in this study was much smaller than that of the bulk sample (approximately a few tens of micrometers) and the thin film (approximately 50 nm), the low mobility can be attributed to the grain boundary scattering of electrons. Moreover, the defects in the films may affect the mobility. The transport properties of nondoped $Mg_2Si$ films were somewhat affected by the substrate temperature, as well as by the sputtering target area ratio. By decreasing the sputtering target area ratio of Mg from 75.8% to 68.2 %, the mobility increased to 5.9 $cm^2/Vs$. When the substrate temperature was increased up to 573 K, the mobility decreased to 0.2 $cm^2/Vs$ because of the generation of small cracks at the grain boundaries in the films. The electron concentrations of Al- and Bi-doped $Mg_2Si$ films were $4.2 \times 10^{19}$ $cm^{-3}$ and $3.9 \times 10^{20}$ $cm^{-3}$, respectively, which is higher by three or four magnitudes than that of nondoped $Mg_2Si$ films. The results clearly show that Al- and Bi-doping into $Mg_2Si$ crystal can be achieved by the co-sputtering of Mg, Si, and Al or Bi at $Ts$ = RT. The electron mobility of Al- and Bi-doped

**Table 1** Transport properties of nondoped and Al- and Bi-doped $Mg_2Si$ films deposited on glass.

| Sample No. | Impurity | Target area ratio | | | $Ts$ | Carrier type | Electrical resistivity | Carrier concentration | Mobility |
|---|---|---|---|---|---|---|---|---|---|
| | | Mg (%) | Si (%) | Impurity (%) | (K) | | ($\Omega$cm) | ($cm^{-3}$) | ($cm^2/Vs$) |
| 1 | – | 75.8 | 24.2 | – | RT | N | 1.42E+02 | 2.2E+16 | 2.0 |
| 2 | – | 68.2 | 31.8 | – | RT | N | 2.05E+01 | 5.2E+16 | 5.9 |
| 3 | – | 75.8 | 24.2 | – | 423 | N | 7.38E+01 | 3.2E+16 | 2.7 |
| 4 | – | 75.8 | 24.2 | – | 473 | N | 6.70E+01 | 5.6E+16 | 1.7 |
| 5 | – | 75.8 | 24.2 | – | 573 | N | 5.41E+02 | 5.7E+16 | 0.2 |
| 6 | Al | 74.5 | 24.2 | 1.3 | RT | N | 8.51E-02 | 4.2E+19 | 1.7 |
| 7 | Bi | 75.5 | 24.2 | 0.3 | RT | N | 6.84E-03 | 3.9E+20 | 2.4 |

$Mg_2Si$ films are 1.7 $cm^2/Vs$ and 2.4 $cm^2/Vs$, respectively, which is almost same as that of nondoped $Mg_2Si$ films.

## CONCLUSIONS

Mg–Si thin films were fabricated on glass, Si(100), Si(111), and polycrystalline $Al_2O_3$ substrates by RF magnetron-sputtering deposition using an elemental composite target composed of Si chips on a Mg disk. Pure-phase $Mg_2Si$ polycrystalline films were successfully fabricated at room temperature. The crystalline orientation of the films was strongly influenced by the Mg/Si elemental composition ratio in the targets, as well as by the surface roughness and porosity of the substrate. The electron concentration and mobility of nondoped $Mg_2Si$ films are $2.2 \times 10^{16}$ $cm^{-3}$ and 2.0 $cm^2/Vs$, respectively. The electron concentration of $Mg_2Si$ films was drastically increased by impurity doping with Al and Bi.

## ACKNOWLEDGMENTS

This research was partially supported by Grants-in-Aid, for Scientific Research (C), No.22560738, 2012 from the Ministry of Education, Sports, and Culture, Science and Technology (MEXT), Japan.

## REFERENCES

1. R. G. Morris, R. D. Redin, and G. C. Danielson, *Phys. Rev.* **109**, 1909 (1958).
2. A. Stella, and D. W. Lynch, *J. Phys. Chem. Solids* **25**, 1253 (1964).
3. W. J. Scouler, *Phys. Rev.* **178**, 1353 (1969).
4. C. B. Vining in *CRC Handbook of Thermoelectrics*, edited by D. M. Rowe (CRC Press, Boca Raton, 1995) pp. 277-285.
5. V. K. Zaitsev, M. I. Fedorov, I. S. Eremin, and E. A. Gurieva in *Thermoelectrics Handbook Macro to Nano*, edited by D. M. Rowe (CRC Press, Boca Raton, 2006) Chapter 29.
6. A. Vantomme, J. E. Mahan, G. Langouche, J. P. Becker, M. V. Bael, K. Temst, and C. V. Haesendonck, *Appl. Phys. Lett.* **70**, 1086 (1997).
7. T. C. Hasapis, E. C. Stefanaki, A. Siozios, E. Hatzikraniotis, G. Vourlias, P. Patsalas, and K. M. Paraskevopoulos in *Energy Harvesting–Recent Advances in Materials, Devices and Applications*, edited by R. Venkatasubramanian, H. B. Radousky, and H. Liang, (Mater. Res. Soc. Proc. **1325**, Cambridge University Press, New York, 2011) pp. 23-28.
8. Q. Xiao, Q. Xie, K. Zhao, Z. Yu, and K. Zhao, *Phys. Procedia* **11**, 130 (2011).
9. T. Yamaguchi, K. Kondoh, T. Sekikawa, M. Henmi, and H. Oginuma, *J. Jpn. Powder Powder Metal.* **52**, 276 (2005).
10. T. Kato, Y. Sago, and H. Fujiwara, *J. Appl. Phys.* **110**, 063723 (2011).
11. H. P. Klug, and L. E. Alexander, *X-Ray Diffraction Methods for Polycrystalline and Amorphous Materials*, (Wiley & Sons, New York, 1974).
12. J. Tani, M. Takahashi, and H. Kido, *J. Alloys. Compd.*, **488**, 346 (2009).
13. K. G. Lyon, G. L. Salinger, C. A. Swenson, and G. K. White, *J. Appl. Phys.* **48**, 865 (1977).
14. Technical data sheet of Matsunami Glass Ind., Ltd.
15. J. Tani, and H. Kido, *Physica B* **364**, 218 (2005).

Mater. Res. Soc. Symp. Proc. Vol. 1490 © 2013 Materials Research Society
DOI: 10.1557/opl.2012.1733

# Enhanced Electrocaloric Effect in Poly(vinylidene fluoride-trifluoroethylene)-based Composites

Xiang-Zhong Chen[1, 2], Xiao-Shi Qian[1], Xinyu Li[1], David Sheng-Guo Lu[1], Haiming Gu[1], Minren Lin[1], Qun-Dong Shen[2], and Qiming Zhang[1]

[1] Materials Research Institute and Department of Electrical Engineering,
The Pennsylvania State University, University Park, PA 16802, USA
[2] Department of Polymer Science & Engineering and Key Laboratory of Mesoscopic Chemistry of MOE, School of Chemistry & Chemical Engineering
Nanjing University, Nanjing, 210093, China

## ABSTRACT

The poly(vinylidene fluoride-trifluoroethylene) (P(VDF-TrFE)) based ferroelectric and relaxor materials have been proved to be good electrocaloric (EC) materials. To further enhance the EC effect in ferroelectric relaxor terpolymer poly(vinylidene fluoride–trifluoroethylene–chlorofluoroethylene) (P(VDF-TrFE-CFE)), composites such as polymer-polymer blends and nanocomposites filled with inorganic nanoparticles are fabricated and investigated. It is found that the addition of small amount of filler (such as P(VDF-TrFE) or nano-$ZrO_2$) can increase terpolymer's crystallinity and enhance its relaxor behavior through interface couplings. The increased crystallinity and enhanced relaxor behavior together result in enhanced electrocaloric effect. The results demonstrate the promise of composite approaches in tailoring and enhancing ECE in the relaxor terpolymers.

## INTRODUCTION

When subject to a change of electric field, the dielectric materials will undergo a polarization change, leading to a change in entropy and thus temperature, which is defined as electrocaloric (EC) effect [1-3]. Compared with the traditional vapor-compression method, cooling with EC effect is more efficient and environmentally friendly. Besides, it may provide efficient means to realize solid state cooling devices for a broad range of applications such as on-chip cooling and temperature regulations for sensors and electronic devices.

Although ECE in ferroelectric materials have been studied for many decades, the relatively small ECE found in bulk ceramics was unimpressive for practical applications. Recently, several groups reported large ECE in ceramic thin films and ferroelectric polymers [4-5], reviving the interests of utilizing these materials for advanced cooling devices. One class of ferroelectric polymers exhibiting large ECE is the P(VDF-TrFE) based relaxor ferroelectric terpolymer. For polymeric materials, it is well known that composites such as blends and nanocomposites can enhance the material properties greatly [6-8]. In this paper, we investigate ECE and associated properties of P(VDF-TrFE-CFE) terpolymer/P(VDF-TrFE) copolymer blends and P(VDF-TrFE-CFE) terpolymer/$ZrO_2$ nanocomposites. It is found that the addition of small amount of filler (P(VDF-TrFE) or nano-$ZrO_2$) can increase terpolymer's relaxor behavior through interface couplings, which results in enhanced electrocaloric effect. Besides, the blends of P(VDF-TrFE-CFE) with small amount of P(VDF-TrFE) also provide a model system to study how the random defects in the terpolymer influence the polarization response in the copolymer. These results demonstrate the promise of composite approaches in tailoring and enhancing ECE in the relaxor terpolymers.

## EXPERIMENT

The relaxor ferroelectric terpolymer P(VDF-TrFE-CFE) 62.5/29/8.5, ferroelectric copolymer P(VDF-TrFE) 55/45 mol % powders were dissolved in N,N-dimethylformamide at room temperature. The $ZrO_2$ nanoparticles (~25 nm) purchased from Alfa Aesar were first surface-treated and then also dispersed in DMF and sonicated for about 0.5 h to break the agglomeration. Then the copolymer solution or $ZrO_2$ dispersion was mixed with terpolymer solution with proper ratio. The mixture was then poured onto a clean glass slide and dried at 50°C. After the solvent evaporated, the composite film was peeled off from the glass and annealed at 105 °C~110 °C for 24h to improve crystallinity and remove residual solvent.

For dielectric measurements, gold electrodes were sputtered on both surfaces of the polymer films. The dielectric properties as a function of temperature were characterized using a precision LCR meter (HP4284A) equipped with a temperature chamber (Delta9023). Polarization-electric field (P-E) loops were measured using a modified Sawyer–Tower circuit at a frequency of 10Hz at room temperature. For ECE measurement, Al electrodes were thermally evaporated on both surfaces of the polymer films. The ECE temperature change was measured using a high sensitivity heat flux sensor in a precise temperature controlled chamber. [9]

## RESULTS AND DISCUSSION

### Enhanced Electrocaloric Effect in terpolymer/copolymer blends

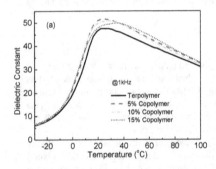 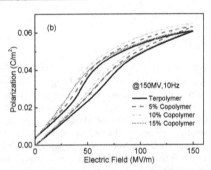

**Figure 1.** Temperature dependence of dielectric properties (a) and unipolar P-E loops of blends (b) of P(VDF-TrFE-CFE) terpolymer and its blends with different amount of P(VDF-TrFE) copolymer.

Fig. 1 (a) shows the temperature-dependent dielectric responses of terpolymer and its blends measured at 1kHz. For the terpolymer there is one diffused peak observed near room temperature. With addition of small amount of the copolymer (5 wt%), the dielectric constants at room temperature increase. The enhancement saturates at 10 wt% copolymer, and then the dielectric

236

constants gradually decrease with further addition of the copolymer. Meanwhile, the maximum of dielectric constant gradually moves towards high temperature. It is interesting that an increase in dielectric constant is obtained by adding a low dielectric constant component (~17 at 1 kHz at room temperature for copolymer P(VDF-TrFE) 55/45). It is believed that the copolymer is converted to relaxor through interfacial couplings, and thus enhances the overall relaxor properties of blends, including the increased dielectric constant. [8]

Enhancement is also observed for the dielectric response at high electric field. Fig. 1 (b) presents the polarization behavior of pure terpolymer and its blends under a unipolar 150 MV/m 10 Hz AC field at room temperature. The blends with 5 wt% and 10 wt% copolymer exhibit higher polarization level than that of pure terpolymer. For example, the polarization of blends with 10 wt% copolymer is 0.066 C/m², 10% higher than 0.06 C/m² of pure terpolymer. Although for blend with 15 wt% copolymer, the enhancement in polarization becomes smaller, the polarization at fields < 150 MV/m is still higher than that of terpolymer.

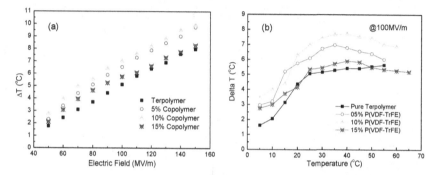

**Figure 2.** Adiabatic temperature change as a function of electric field at room temperature (a) and as a function of temperature at 100 MV/m (b).

Fig. 2(a) presents the adiabatic temperature change (ΔT) of the terpolymer and its blends as a function of electric field measured at room temperature. As can be seen, blends show higher ΔT than that of the neat terpolymer at all electric fields measured. For example, at 150 MV/m, a ΔT ~ 8 °C is induced in pure terpolymer, while in blends with 10 wt% copolymer a ΔT ~ 10 °C is obtained. For blends with 15 wt% of copolymer, ΔT is higher at fields < 100 MV/m and above that ΔT is nearly the same as that of pure terpolymer.

The ΔT as a function of sample temperature under 100 MV/m is shown in Fig. 2(b). Above room temperature, ΔT displays a weak temperature variation, whose trend is similar to the temperature variation of the dielectric response observed in terpolymer and its blends. That is, the ΔT peak shifts towards higher temperature with increased copolymer content, in analogous to the dielectric behavior (Fig. 1a). These results indicate that the dielectric responses observed in the blends affect directly the electrocaloric properties. As temperature is lowered toward the freezing temperature of the relaxor, remnant polarization will develope after application of high electric fields, which will result in a decrease of ΔT.

## Enhanced Electrocaloric Effect in terpolymer based nanocomposites

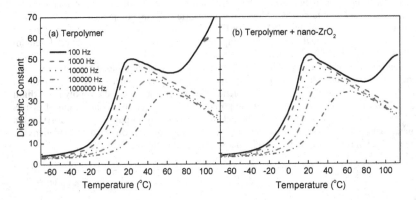

**Figure 3.** Temperature dependence of dielectric properties of P(VDF-TrFE-CFE) terpolymer (a) and terpolymer-based nano composites (b).

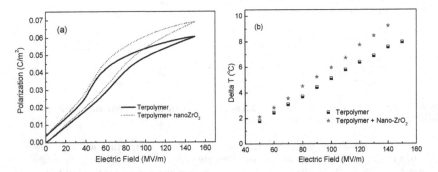

**Figure 4.** (a) Typical unipolar P-E Loops of terpolymer and nanocomposite measured at room temperature at 150MV/m. (b) The adiabatic temperature change of terpolymer and nanocomposite measured at room temperature under different electric field.

Fig. 3 presents the weak field dielectric properties of the terpolymer and its nanocomposite. One thing worth noticing is around room temperature, a little bump appears in the dielectric constants of nanocomposites, especially at low frequencies, which may be caused by the contribution of interface areas. Another noticeable effect is for the dielectric properties measured at 100 Hz and above 80 °C, where the space charge contribution (conduction) causes a large increase of the dielectric constant (Maxwell-Wagner space charge effect) with temperature in the terpolymer, whereas the nanocomposite shows reduced loss and dielectric constant, indicating that the introduction of $ZrO_2$ nano-particles reduces the conduction loss, presumably due to an increased trap density in the nanocomposites.

Fig. 4(a) presents the polarization behavior of pure terpolymer and nanocomposites with 3% vol $ZrO_2$ measured under a unipolar 150 MV/m field at 10 Hz and room temperature. The nanocomposites exhibit 15% higher maximum polarization (0.069 $C/m^2$) than that of terpolymer. The small remnant polarization is same as terpolymer, indicating the enhancement of maximum polarization is intrinsic, and is not due to conduction loss. Fig. 4(b) compares the adiabatic temperature change ($\Delta T$) of the terpolymer and nanocomposite as a function of electric field measured at room temperature. As can be seen, nanocomposite shows higher $\Delta T$ than that of the neat terpolymer at all electric fields measured. For example, at 140 MV/m, a $\Delta T \sim 9.2$ °C is induced in nanocomposite, about 20% higher than pure terpolymer ($\sim$7.6 °C). To make quantitative comparison, we note that based on symmetry consideration, entropy change $\Delta S$ in a dielectrics is related to the electric displacement D through $\Delta S = - \beta D^2/2$, where $\beta$ is a coefficient. Applying $c_E \Delta T = -T\Delta S$ yields the adiabatic temperature change $\Delta T = \beta D^2/(2Tc_E)$. Within the experimental error, $\beta$ does not show marked change with blend composition, indicating that the enhancement in ECE in nanocomposite is a direct consequence of the enhanced polarization response.

## CONCLUSIONS

In conclusion, ferroelectric relaxor terpolymer poly(vinylidene fluoride-trifluoroethylene-chlorofluoroethylene) (P(VDF-TrFE-CFE)) based composites such as polymer-polymer blends and nanocomposites are fabricated and investigated. It is found that the addition of small amount of filler (such as P(VDF-TrFE) or nano-$ZrO_2$) can enhance its relaxor behavior through interface couplings which results in enhanced electrocaloric effect. These results demonstrate the promise of composite approaches in tailoring and enhancing ECE in the relaxor terpolymers.

## ACKNOWLEDGMENTS

This research was supported by Army Research Office under Grant No. W911NF-11-1-0534 (X. Chen, S.G. Lu, M. Lin, and Q. Zhang) and by the U.S. DoE, Office of Basic Energy Sciences, Division of Materials Science and Engineering under AwardNo.DE-FG02-07ER46410 (X. Qian and X. Li). X. Chen was also partially supported by Nanjing University, China and H. Gu was in part supported by EE Department of PSU.

## REFERENCES

1. E. Fatuzzo and W. J. Merz, *Ferroelectricity.* (North-Holland Publishing Company, Amsterdam, 1967).
2. T. Mitsui, I. Tatsuzaki and E. Nakamura, *An Introduction to the Physics of Ferroelectrics.* (Gordon and Breach, London, 1976).
3. M. E. Lines and A. M. Glass, *Principles and Applications of Ferroelectrics and Related Materials.* (Clarendon Press, Oxford, 1977).
4. A. S. Mischenko, Q. Zhang, J. F. Scott, R. W. Whatmore and N. D. Mathur, Science **311**, 1270 (2006).
5. B. Neese, B. J. Chu, S. G. Lu, Y. Wang, E. Furman and Q. M. Zhang, Science **321**, 821 (2008).

6. B. J. Chu, M. R. Lin, B. Neese, X. Zhou, Q. Chen and Q. M. Zhang, Appl. Phys. Lett. 91, 122909 (2007).
7. J. J. Li, S. I. Seok, B. J. Chu, F. Dogan, Q. M. Zhang and Q. Wang, Adv. Mater. 21, 217 (2009).
8. X. Z. Chen, X. S. Qian, X. Y. Li, S. G. Lu, H. M. Gu, M. R. Lin, Q. D. Shen, Q. M. Zhang, *Appl. Phys. Lett.* 100, 222902 (2012).
9. X. Y. Li, X. S. Qian, S. G. Lu, J. P. Cheng, Z. Fang and Q. M. Zhang, Appl. Phys. Lett. 99, 052907 (2011).

# AUTHOR INDEX

# SUBJECT INDEX

Printed in the United States
by Baker & Taylor Publisher Services